MEASURING
THE COSMOS

MEASURING THE COSMOS

How Scientists Discovered
the Dimensions of the Universe

DAVID H. CLARK *and*
MATTHEW D. H. CLARK

RUTGERS UNIVERSITY PRESS
New Brunswick, New Jersey,
and London

Library of Congress Cataloging-in-Publication Data

Measuring the cosmos : how scientists discovered the dimensions of the
universe / David H. Clark and Matthew D.H. Clark.
p.   cm.
Includes bibliographical references and index.
ISBN 0-8135-3404-6 (hardcover : alk. paper)
1. Cosmology—Research.   I. Clark, David H.
II. Clark, Matthew D. H., 1970–
QB981.M385 2004
523.1—dc22
2003020096

A British Cataloging-in-Publication record for this book is available
from the British Library.

Manufactured in the United States of America

To David Samuel, Marcus Alexander, and Peter James

CONTENTS

PREFACE AND ACKNOWLEDGMENTS

This book tells the stories of "heroes" and "heroines," both well known and unsung, behind the present understanding of the origin, size, and age of the cosmos. Edwin Hubble, Albert Einstein, Fred Hoyle—these are household names. But how many people, other than professional astronomers, have heard of scientists like Heber Curtis, Henrietta Swan Leavitt, Annie Jump Cannon, Milton Humason, or Vesto Slipher? Our current understanding of the nature of the universe depends on all of them, not just the famous few.

Nor was the path to contemporary knowledge a well-ordered progression from ignorance to understanding. Along the way, astronomers, mathematicians, and physicists made mistakes, faced impenetrable uncertainties, and found plenty of time to fight within their ranks. The stories of the scientists involved in measuring the cosmos reveal ambitions, conflicts, failures, as well as successes as the astonishing scale and age of the universe were finally established. Few areas of scientific research have witnessed the likes of the clashes of egos, the claims and counterclaims of priority of thought, or the failed (or falsified) theories and observations that resulted from attempts to measure the universe. The ancient Greeks used the word cosmos to describe the observable heavens. The word cosmos means order and hierarchy (and is the opposite of chaos). Not only the universe, but also the emergence of a true understanding of its complexity, would prove to be more chaotic than the Greek scholars could ever have imagined for a system they believed to be perfect and well ordered.

PREFACE AND ACKNOWLEDGMENTS

The story that follows gives overdue credit to some of the little-known individuals who played a key role in the race to measure the cosmos.

We wish to thank our agent, Al Zuckerman of Writers House, and our editor, Audra Wolfe, for their helpful guidance and advice. The suggestion to write a book on measuring the cosmos came from John Michel, and we are grateful for his guidance during the formative stage of the project. Suzanne Clark and Agnes Clark provided much valued research assistance. We are enormously grateful to all the members of our family for their support and tolerance during the long hours of writing.

David Clark and Matthew Clark, July 2003

MEASURING
THE COSMOS

Prologue

A RAPID JOURNEY THROUGH
TIME AND SPACE

Humans have always viewed the heavens with wonder and with awe, sensing, as they looked out into the night sky, the vastness of space, the power of the creation, and perhaps even something of their own origins. The importance of the heavens in the lives of primitive peoples is demonstrated by depictions of the Sun and stars in cave paintings and on ancient monuments, the emergence of Sun and Moon worship, the development of simple calendars based on the changing patterns of the skies, the use of the stars for navigation, and the development of astrology.

Classical civilizations sought an improved understanding of the changing patterns in the heavens through logic. Thales of Miletus in the sixth century B.C.E. was the first to explain natural phenomena through philosophical reason, correctly predicting the occurrence of a solar eclipse. Anaximander, one of Thales' most noted successors from Miletus, introduced the notion of the "infinite"—a universe that was infinite in time and space, with things being brought into being and passing away. In modern times Anaximander's notion would reemerge as the so-called steady state theory of the universe. By contrast the philosopher Anaxagoras believed that at some time "all things were together," an idea that would be presented in the twentieth century as the big bang theory for the origin of the universe. Thus the ancient forebears of modern cosmologists were already grappling with the issues of the scale and origin of the cosmos.

Aristotle forged the notion of a perfect universe centered on

1

the Earth. Prior to Aristotle it was argued that everything of common experience could be explained in terms of the four "elements": earth, fire, air, and water. Aristotle introduced a fifth perfect "element," the so-called quintessence, to explain what he perceived as the perfect nature of things in the heavens. The last of the great Greek astronomers was Claudius Ptolemy. His famous work *The Great Synthesis*, commonly called the *Almagest*, presented formally the accepted astronomical theories of the day based on a universe that had the Earth at its center around which the Sun, planets, and stars revolved. The impact of the *Almagest* was truly astounding. The Earth-centered model of the universe remained essentially intact through the dark ages of intellectual stagnation following the collapse of the Greco-Roman culture. By the twelfth century Ptolemy's teachings, based on thinking of a millennium earlier, formed a cornerstone of the religious dogma of the Church of Rome. Questioning the Ptolemaic doctrine became tantamount to heresy.

The first serious alternative to the Ptolemaic theory appeared when the Polish cleric Nicolaus Copernicus published his theory of a Sun-centered system in the monumental treatise *De revolutionibus orbium coelestium* (On the Revolution of the Heavenly Bodies) shortly before he died in 1543. Copernicus proposed that the Earth and other planets orbited the Sun rather than everything in the heavens orbiting the Earth. In fact the Greek philosopher Aristarchus had argued for a heliocentric cosmos in the third century B.C.E., but his thesis had not prevailed. Copernicus's ideas were to rock human understanding and challenge the infallibility of papal doctrine in the century to follow. The very word revolution, when applied to the violent overthrow of an existing system, has its origins in the treatise.

Although Copernicus initiated the revolution in human understanding of the universe, it has been only through the scientific advances of the past 150 years that people could gain a real appreciation of its true enormousness, its cataclysmic origin billions of years ago, or their own close relationship to the stars. The historical picture, dominated in the West initially by the philosophies of Greek scholars and subsequently by Christian teachings, changed dramatically as scientists finally learned how to estimate the incredible distances to cosmic objects with some certainty. As a consequence we are now closer to a true appreciation of the vastness of the universe, its cataclysmic origin, and its likely fate. The vision of the heavens imagined by some of the ancients, an infinite cosmos populated with a myriad of Sun-like stars, turns out to be closer to reality than the overly constrained view that pertained for almost two thousand years up to the past century. The scope of present-day investigation extends from speculation about the origin of the universe, to studying its present turbulent state, to thinking about its ultimate fate. It extends from Earth's nearest and comparatively well-understood stellar neighbors to bizarre and enigmatic objects at the extremities of an observable universe that is in a state of constant and violent upheaval.

It might be considered a somewhat strange detective novel that revealed briefly in its early pages the resolution of a murder mystery, before then proceeding to describe the nature of the crime and the gathering of evidence needed to solve it. Nevertheless that is the approach to be adopted here. We will start by describing the scale and age of the cosmos as revealed by modern research, before looking in the chapters that follow at how scientists developed an appreciation of its true enormity by establishing methods to measure the distances to the stars and galaxies, and thence estimating the likely age of the universe.

Stars are not uniformly scattered throughout the cosmos but accumulate in vast conglomerates called galaxies. Our Sun is just one of an estimated 400 billion stars within our galaxy, which since antiquity has been called the Milky Way because of its white cloudy appearance running across the night sky. The Milky Way turns out to be discus shaped (we are seeing it from within), with our Sun occupying a rather insignificant location closer to its periphery than to its heart. The Milky Way appears at its most spectacular when viewed from Earth's Southern Hemisphere, where one is looking toward the Galaxy's center. Within the Milky Way we find clusters of millions of stars closely packed together, embedded in a more systematic distribution of stars.

The German philosopher Immanuel Kant, drawing on an idea of the British astronomer Thomas Wright, speculated in his work *Universal Natural History and Theory of the Heavens* in 1775 that there were numerous other Milky Ways—accumulations of stars that he referred to as "island universes." However, his idea was largely ignored, and until 1924 many scientists believed that the universe did not extend beyond the Milky Way. How mistaken our scientific forebears were! If our place within the Milky Way seems insignificant, then it is now appreciated that within the overall universe the place of the Milky Way itself is similarly insignificant. Modern astronomical techniques have now revealed that the observable universe contains vast numbers of galaxies.

It was one of the great feats of astronomy to learn how to determine the distances of the stars. Now that we have some idea of the universe's enormous size, we find that familiar units of distance—such as mile or kilometer—don't begin to capture its scale. Instead we use a unit of distance called the light-year, the distance a pulse of light travels in one year. Since the speed of light is a staggering 300,000 kilometers each second, a light-year is a considerable distance. The Sun is eight light-minutes away; sunlight illuminating Earth left the Sun eight minutes ago. The outer planets are light-hours away. We now know that the nearest star to the Sun is about five light-years away, while the more distant stars we see in the night sky with the naked eye are many thousands of light-years away.

The Milky Way is 100,000 light-years across. Nearby galaxies are then millions of light-years distant. These can sometime be found in "groups" several million light-years across containing dozens of galaxies, and there are "clusters" tens of millions of light-years across that contain thousands of individual galaxies. These all form "superclusters" hundreds of millions of light-years across. The clusters and superclusters have a connecting structure like tangled spaghetti around giant voids, or perhaps like a Swiss cheese. The universe probably contains on the order of 100 billion galaxies. Current understanding suggests that very distant galaxies are at distances of billions of light-years. On the universal scale, planet Earth must be considered to be no more than a mere speck of cosmic sand swept by the tides of universal change.

The process for estimating cosmic distances involves a series of steps and methods as we move out from planet Earth. Imagine that we have a ruler marked in inches and feet for measuring features around the home. We then use this ruler to calibrate one marked in feet and yards to measure room sizes—and then we use the yard ruler to calibrate to one marked in tens of yards to measure distances around the garden—and thence continue to one marked in miles to measure distances around the countryside. If we did not have the inches and feet on our first ruler accurately marked, then the subsequent calibration of rulers for larger scales would be wrong. It is the same with measuring cosmic distances; as we move farther and farther out into the cosmos using complementary techniques to estimate distance, it is important to check the calibration of overlaying methods of measurement. Although there has been ample investment of intellectual effort and generous funding, the task is not easy and has required a monumental effort on the part of the scientists involved. Many of the surveys required years of observation, and in some cases researchers devoted their whole careers to trying to perfect a single distance-measuring technique.

The first step in measuring cosmic distances is by the process known as parallax. A simple illustration of parallax is to hold a finger upright at arm's length. First view the finger with your left eye closed—and then with your right eye closed. The finger will appear to have moved with respect to background objects, despite its having been held stationary. Similarly if a star fairly close to Earth is observed six months apart, when Earth is on opposite sides of its orbit around the Sun, then the star will appear to have shifted position with respect to more distant "background" stars. Copernicus and others realized that a consequence of his theory that Earth orbited the Sun must be that stars would display parallax. The fact that parallax could not be detected with the best measuring instruments they had available implied that the stars must be at unimaginable distances.

Return for a moment to the experiment with a finger held at arm's length, and imagine that with your right eye closed the finger is aligned exactly with a distant object. Now with your left eye closed it is no longer aligned with the distant object, since it is being viewed from a slightly different position. The angle of displacement can be estimated. If you know the distance between your eyes, you can use a method known as triangulation to estimate the distance to your finger. Since antiquity, triangulation has been the standard method used by surveyors to estimate the distance to remote objects. First the direction to a remote object is determined from one position. (This can be the compass direction.) Then the direction is determined from another position a known distance from the first. Triangulation then gives the distance to the remote object using the difference in direction (an angle), the distance between the two places of observation, and simple trigonometry.

Triangulation can, for example, be used to estimate the distance to the Moon. Suppose the position of the Moon with respect to the background stars is measured at two points widely spaced on Earth's surface at a prearranged time. By knowing the distance between the two observing points, one can triangulate to get the distance to the Moon as about thirty times Earth's diameter.

The first person to successfully measure the distance to a star from its parallax was Friedrich Bessel, the director of the Konigsberg Observatory in Germany. Bessel started his search for stars exhibiting parallax in 1837 with a binary star system in the constellation of Cygnus. He calculated its distance to be eleven light-years, considered a staggering distance at the time.

By 1878 only seventeen stars had parallax distances determined; such was the difficulty in making the measurements. Indeed, even by 1900 fewer than one hundred stellar distance measurements had been obtained. But in 1903 Frank Schlesinger introduced the technique of photography to parallax measurement, enabling the method to be extended to estimating the distances of the several thousand stars closest to the Sun. The distances determined, many tens of light-years, were now extending the bounds of human understanding. But this was merely a modest first step!

In looking out into the cosmos through a telescope, one is looking not only deep into space but also back in time. Thus the nearby stars are viewed as they were several years ago, and more distant stars within the Milky Way as they were thousands of years ago, when the light now reaching Earth began its cosmic journey. The nearby galaxies appear as they were millions of years ago, and the more distant galaxies as they were hundreds of millions or even billions of years ago. Few of the objects we now observe still exist (at this instant) in the form we presently see them, and some may no longer even exist. Thus the history of the universe is laid out for Earthbound heaven gazers to contemplate. The telescope represents a "time machine" in which we can study stars and galaxies at various stages of their evolution: nascent stars procreated from giant clouds of interstellar gas and dust; young stars, old stars, dying stars, and dead stars; young galaxies, interacting galaxies, and galaxies being torn apart. The universe reveals itself to be a spectacle of unfolding drama as stars and star systems are born and die, often violently.

We must adjust our terrestrial scale of thinking if we are to appreciate the masses and timescales involved in describing the universe. We choose to measure the mass of objects of common experience in terms of a convenient standard mass, the kilogram. Thus, for example, an adult male may have a mass of about 80 kilograms. The mass of planet Earth is 6 million billion billion kilograms! The Sun is some 300,000 times more massive than Earth. The mass of the Milky Way is probably at least 100 billion times that of the Sun! And the mass of the universe is certainly greater (perhaps much greater) than 1,000 billion billion solar masses. No less of a challenge to the human imagination are the timescales involved in describing astronomical phenomena. Earthbound events are conveniently measured in terms of the sidereal year; the time for Earth to complete one orbit about the Sun, measured relative to the fixed stars. The Sun and other stars orbit around the center of the Milky

Way. At the Sun's distance from the Milky Way's center, the stars take over 200 million years to complete one revolution. The Sun is believed to be some 5 billion years old; and will survive for a similar period. The Milky Way is probably some 8 to 10 billion years old. The universe itself is thought to be at least some 13 to 14 billions years old.

Such vast times and distances would have shocked astronomers from the ancient and early modern worlds. Building on the traditions of ancient Greeks and Babylonians, Renaissance and early modern astronomers developed a complex set of theories about the nature of the universe. Their ideas about the relationship of the Earth to the heavens had important consequences for religion, politics, commerce, and exploration. Our story therefore rightly begins with a survey of astronomical knowledge from the sixth century B.C.E. to the mid–nineteenth century. But despite the dramatic advances made in over two millennia of scientific observation, nineteenth-century astronomers still lacked concrete evidence for the true dimensions of the cosmos.

Since the nineteenth century, science has moved in just one hundred years from a position of almost total ignorance about the actual distances to the stars, and about the nature and distances of galaxies, to our present knowledge of the enormous size, mass, and age of the universe. We are reaching the limits of observation, and therefore the limits of human understanding. Beyond lies only our imagination, seeded by the inspiration of the theories of physics. But the race to measure the cosmos goes on.

Chapter 1

INGENIOUS VISIONS

Astronomy owes its heritage
to many historical strands, foremost among them being the studies
of the heavens by the ancient Babylonians, the Chinese and other
Far Eastern cultures, the early Egyptians, and the Mesoamerican
civilizations. It was the ancient Greeks, however, who set in place
a vision of the cosmos that influenced European thinking until the
Renaissance. The ancient Greek thinkers are the natural ancestors
of current cosmologists, hence the attention that we give to them
here. Their studies would impact human understanding of the size
and nature of the cosmos for almost two millennia, and their philosophy and observational creativity remain a source of fascination.

Ancient Greek tradition saw the Earth as a flat disk, with the
heavens as a dome encompassing the disk. The stars were fixed to
the dome. But some celestial objects were found to wander among
the fixed stars: they were called the planets (the "wandering stars").

The ancient cosmological worldview was dominated by beliefs
about the gods. This applied especially to the great "why?" questions. When any wise man or woman of an ancient tribe was asked
a question such as "Why does the rain fall?" he or she would answer with some myth—perhaps explaining how the rain was the
result of the will of the sky god and his gift to humankind. Myths
were handed down from generation to generation, although there
was ample freedom for the embellishment of old myths and the
creation of new ones. Modern anthropologists believe that humanity began using such myths at the time of the ancient cave paintings, and that these myths were the motivation behind the development of art and much else of ancient culture. Such "why" myths

9

have had a great influence on human development, but their fundamental weakness was that they lacked the power of prediction. A solar eclipse may be explained as the grief of the Sun god. However, this makes the phenomenon unpredictable—and remote from human experience. In 585 B.C.E. the Greek astronomer Thales of Miletus forever changed human experience by correctly predicting the date of a solar eclipse. Thus the ancient Greeks were presented with two different explanations of this event, one mythological and the other scientific. While the mythmakers had centuries of tradition on their side, only Thales had the power of prediction.

Greek astronomical research dates back well before Thales, and Thales himself must have relied on calculations produced by the Babylonians, who were gifted observers of the heavens. In fact Thales of Miletus got very lucky. If his calculations were indeed based on the Babylonian method, he would have been able to predict accurately lunar, but not solar, eclipses. He was particularly fortunate that the eclipse of 585 B.C.E. was total as viewed from the scene of a reasonably significant battle between the Lydians and the Persians. And he was lucky that the first Greek historian, Herodotus, recorded his prediction for prosperity.

> The war [between the Lydians and the Persians] was equally balanced, until in the sixth year an engagement took place in which, after battle had been joined, the day suddenly turned to night. This change in the day had been foretold to the Ionians by Thales of Miletus, who had fixed as its term the very year in which it actually occurred.

Despite his good fortune, only the mean-spirited would wish to rob Thales of his scientific immortality. His dramatic prediction demonstrated the value of systematic measurement of the heavens, and it places Thales at the beginning of the story of astronomical prediction.

Another great contribution that Thales made to the history of science is that he was the first Greek philosopher that we know to have referred to the concept of "an element." Certainly he is at the start of the tradition of the search for the underlying substance of all matter, which led in time to the practice of alchemy in the Middle Ages—and would lead to the production of the periodic table when the falsehoods of alchemy were eventually replaced by the scientific method. Thales' imagination led him to propose that the underlying substance of the whole cosmos was water. He believed that the Earth had been produced by the condensation of water and the air had been produced from water by rarefaction.

Thales was a citizen of the city of Miletus, and one estimate of his life is 624–547 B.C.E. In fact, the first three famous Greek philosophers, Thales, Anaximander, and Anaximenes, all came from Miletus, and the location and circumstances of the city were very significant to the birth of some of the fundamental concepts of science.

Miletus was in modern-day Turkey, on the coast of the Aegean Sea. Not far north was Ephesus, home of the Temple of Artemis, one of the seven wonders of the ancient world. Within a few miles to the south was Didyma, the site of one of the major oracles of the Greek world, and farther south was Halicarnassus, the home city of Herodotus, the first historian. The area was a crucible of philosophical reasoning.

What was particularly important about Miletus, from the point of view of our story, is that the city was open to ideas from the East, and in particular from Babylon. Although Babylon has been remembered in folklore for its immorality (based on its widespread ritual prostitution in honor of the goddess Ishtar), it should perhaps be more charitably remembered for its contributions to the origins of astronomy. Hundreds of years before Thales, the Babylonians had plotted the background stars to the setting Sun. Clearly identifiable groups of stars were known as constellations, and myths and legends were assigned to them. The constellations through which the path of the Sun passed were noted. This path defined what was called the zodiac, which the Babylonians divided into twelve "signs" of 30 degrees each. The signs of the zodiac took on particular significance in prognostication. It supposedly made a difference to the ancient diviners which sign of the zodiac one was born under (an ancient myth that is fed to a gullible public to this day by the tabloid press). The Roman historian Pliny records that Cleo-stratus of Tenedos, a Greek, recognized the signs of the zodiac. Modern scholars have placed Cleostratus to the second half of the sixth century B.C.E. and have speculated that he learned from the Babylonians about the signs of the zodiac and about some of the constellations the Babylonians had defined by star patterns. The Babylonians had also discovered the lunar cycle of 223 lunar months, which Thales must have learned about and used as the ba-sis of his eclipse prediction.

The Babylonians might have been able to describe "what" and predict "when." But Thales' achievement, with his fellow Greek philosophers, was to ask the question "why?" It was the transition in reason from "what" and "when" to "why" that set the Greek cultures above all others—even before their contributions to art, music, literature, democracy, and architecture are acknowledged.

Although we have no evidence for Thales' view on the size of the cosmos, we do know that he applied his view about what the cosmos was made of to its structure. In *On the Heavens*, Aristotle attributes to Thales the view that the Earth floats on water like a log in a stream. Simplicius, a much later commentator on Aristotle (who, writing in the sixth century C.E., is one of the most im-portant sources of information on the early Greek cosmologists), suggested that Thales had derived his cosmological beliefs from knowledge of ancient Egyptian mythology. This has received some support from modern scholars of Egyptian beliefs. Certainly Thales' beliefs fit into the wider pattern of cosmological ideas put forward in this period by the civilizations of the Middle East, such as the Hebrews and the Babylonians. However, Thales inspired succes-sors, who continued to speculate about the visible world based on their own observations. It was his belief in the possibility of rational explanation of complex visible phenomena that makes Thales such a giant in the history of natural science.

Of all the ancient natural philosophers, Anaximander is the one who most naturally falls into a chapter titled "Ingenious Visions." Anaximander also lived in Miletus, and the historical tradition re-cords that he was a pupil of Thales. Certainly there are obvious links between the ideas attributed to both men. The dates suggested for Anaximander's life are 611–546 B.C.E. Unlike Thales, Anaximander is recorded as having written a book, titled *On Nature*. Book writing was rare even by the time of Aristotle and Theophrastus in the fourth century B.C.E. Tragically Anaximander's work, although referenced much later, was subsequently lost. Anaximander appears to have written many cosmological theories in his book *On Nature*, but we have to rely on secondary sources, such as the Christian apologist and historian of philosophy Hippolytus (180–235 C.E.), to reconstruct them.

Like Thales, who believed that water formed the underlying substance of the whole cosmos, Anaximander speculated on the ultimate form of matter. In fact the much later historian of philosophy Simplicius mentions the tradition that Anaximander was the first writer to use the word *arche* (to describe the underlying element). However, rather than identifying any common substance such as water as the underlying element or principle, Anaximander argued that the basic material was the "infinite," or the unlimited. He seems to have been struck by the thought that the underlying element must be infinite, so that the processes of coming into being and passing away would be infinite. Plato applies Anaximander's concept of the infinite to souls in the *Phaedo* to argue for the immortality of the soul.

Since Anaximander believed that the underlying principle was infinite, it is natural to think that he believed the size of the universe to be potentially infinite. (Over two millennia later cosmologists would still be grappling with the concept of an "infinite" universe.) Certainly Anaximander's vision of the cosmos and the heavenly bodies was dramatic and extensive. Of all the ancient thinkers, he seems to have come the nearest to imagining the vast size of the universe.

Hippolytus gives a detailed account of Anaximander's ingenious celestial system. The first principle of things is the infinite nature, from which the heavens and Earth are created. Just as the principle substance had to be infinite, so he believed that there had to be eternal motion to explain the balancing process of coming into being and passing away.

Anaximander developed an imaginative theory to explain the

creation of the Earth, Sun, Moon, and stars. He believed that part of the infinite became separated into two opposites, the hot and the cold. The hot section developed into a sphere of flame. Various portions then broke off and were surrounded by rings of compressed air. These rings contained "breathing holes," which allowed the emission of fire and light. Eclipses occurred when these breathing holes became blocked. The imagination behind these speculations was as breathtaking for the ancient Greeks as it appears to be simplistic and nonsensical to us. However, Anaximander appears to be the first thinker to analyze the objects we observe in the sky in natural terms, and to try to explain in a systematic way how we see them—even if to our modern way of thinking his speculations appear somewhat picturesque.

Anaximander was not afraid to jettison the theories of his mentor, Thales. Instead of Thales' theory of the Earth floating in water, he believed that the Earth was not supported by any physical body but remained in place because it was equidistant from all other heavenly bodies. It is natural to assume that this argument presupposed some primitive theory of gravity, whereby the cosmic objects exerted balancing forces on the Earth. He believed that the Earth itself was a short cylinder and that the depth of the cylinder was one-third of its breadth. Living things existed on one of the faces of the cylinder. He also speculated about the distance between the Earth and the observable cosmic bodies. This is the earliest estimate we know about of the distances within the cosmos. He calculated the Moon's distance to be nineteen times the radius of the plane face of the Earth and the Sun's distance to be twenty-eight times the same radius (the former is short by a factor three, and the latter is short by a spectacular margin).

After all this theory, it is a relief to hear that Anaximander had some practical accomplishments, although we know almost nothing about his life. He was recorded as the first person to attempt to draw a map of the whole Earth. Hecateus, also of Miletus, later revised this map, and we know that by the time of the Ionian revolt in 499–493 B.C.E. mapmaking was well advanced in Greek Asia Minor. He is also recorded to have traveled to Sparta and to have set up a sundial recording the solstices, the time, the seasons, and the equinox.

Of course the natural modern response to Anaximander's wonderful visions is to ask what evidence he based these visions on, and of course we do not know. Present sources are summaries of summaries, aiming merely to list his beliefs concisely and dramatically. The arguments and fragments of argument attributed to Anaximander that we do have seem to suggest that he was more inclined than Thales to reason from first principles, rather than to base his beliefs merely on observation. What is indisputably important about his speculation, however, is that he tried to explain what he saw in the heavens and on Earth purely in natural terms rather than assigning causation to the gods. And there are uncanny echoes from some of Anaximander's theories in those which have evolved in modern science, such as the creation of the solar system from a collapsing cloud of hot gas ("ball of fire"), the continual balancing processes of coming into being and passing away applied to cosmic objects, and the theory of balancing attracting forces between large bodies (gravity). These must be seen as real triumphs of the human intellect, even if generations of scientists would be needed before picturesque speculations could be turned into valid scientific hypotheses.

The third great natural philosopher from Miletus is Anaximenes, who lived from around 585 to 528 B.C.E. He was an associate of Anaximander, and thus the line of intellectual succession is continued. Anaximenes' astronomical theories, and indeed his general line of thought, seem closely related to Anaximander's ideas. Diogenes Laertius, writing in the third century C.E., contrasts Anaximenes' "simple and economical Ionian style" with Anaximander's somewhat poetical words. This suggests that Anaximenes' writings survived to the Christian period but not much beyond. Aristotle discusses Anaximenes' astronomical views, and our evidence is supplemented by the Christian writer Hippolytus, writing in the third century C.E.

Anaximenes continued the same line of inquiry as Thales, wishing to discover the primary element. However, he differed from the earlier Milesian natural philosopher by believing that the source of all material objects was air. Like the other two Milesians, he developed astronomical theories, inspired by his belief in his particular element. He argued that the Sun, Moon, and stars originally evolved from the Earth. His theory of the solar system continued by arguing that moisture rose from the Earth and, when rarefied, produced fire. The stars and the other heavenly bodies were all made of fire and rested on air, on account of their breadth and flatness, he suggested. He further believed that the stars were fastened to a crystal sphere, like nails or studs, whereas the planets rode on the air independently. (The notion of "crystal spheres" would survive until the Renaissance!)

Anaximenes also seems to have believed that in the region of the heavens occupied by the stars there were separate bodies like the Earth, which the stars carried round with them. It has been suggested that Anaximenes invented these bodies to explain solar and lunar eclipses—although an "alternative worlds" interpretation is also possible. Thus Anaximenes continued the noble tradition of Thales in seeking to explain celestial phenomena, rather than simply being content with the ability to predict their behavior. A fragment recorded by Aetius sums up Anaximenes' ethereal worldview: "Just as our soul, being air, holds us together, so do breath and air encompass the whole world."

Anaximenes is the last of the line of the great Milesian natural philosophers, who were so significant in the beginning of the history of cosmology and natural science. Unlike his predecessors, Anaximenes would have spent his mature years under the Persian Empire. Herodotus records how Croesus, king of the Lydians, was encouraged by the Delphic oracle's prediction to attack the Persians in 546 B.C.E. The oracle had said that if he attacked the Persians, he would destroy a mighty empire. He happily went on the offensive, only to find in defeat that the mighty empire destroyed was his own! Miletus thus passed from one foreign ruler to another, and it continued to flourish for another fifty years before becoming the center of an Ionian revolt against the Persian satraps. The revolt failed, as the Persians brought in reinforcements from their huge empire. In 494 B.C.E. Miletus was sacked by the Persians and lost its leading role as the major Greek scientific center. That distinction was to pass to the city of Athens, which inflicted two major defeats on the Persians, at Marathon in 490 B.C.E. and at Salamis in 480 B.C.E. But as the baton of enlightenment was passed on, the role of the Milesians should not be forgotten in shaping all that followed.

We will examine the great thinkers of classical Athens later in the chapter. Now we move from the western shores of Asia Minor to the southern shores of Italy. Greek colonists built cities all over the Mediterranean coastline, and few of these colonists are more renowned than Pythagoras. However, it is more accurate to talk about the school of Pythagoras, since many of the discoveries and theories that carry his name are more likely to have been developed by his followers. Pythagorean philosophy ranged from the sublime (for example a veneration for the study of numbers) to the ridiculous (an abhorrence to eating beans). But the Pythagoreans certainly represent the next stage in the history of Greek speculation on the cosmos.

Pythagoras and his early followers left no written works that survived. Their ideas endured through the traditions of their oral teaching, which were later recorded by other philosophers. Plato and Aristotle refer to Pythagorean teaching, and Plato was clearly highly influenced by Pythagorean arguments for the immortality of the soul. It is very difficult to date these arguments, however, and especially difficult to know which arguments and ideas actually date back to Pythagoras himself. In the fourth century B.C.E. and afterward, the histories of Pythagoreanism and Platonism became interlinked, and thus any interpretation of their respective contributions is bound to be somewhat confused.

Despite the obscurity of his personal ideas, there is a rich source of anecdotes about Pythagoras, and later historians recorded a number of details of his eventful and turbulent life. He was born around 570 B.C.E. on the island of Samos, not far from Miletus, and thus it is plausible to believe that he was aware of the ideas of the Milesian natural philosophers. Around 540 B.C.E. he emigrated from Samos to the distant Greek colony of Croton in southern Italy. He founded his first school in Croton, teaching both men and women, and he became very prominent in the city. Later, however, he was exiled from Croton, probably because of political or religious controversy, and he settled in the nearby southern Italian town of Metapontum, which was also a Greek colony. Here he continued to lead his school, and it is likely that many of his students shared in his exile, during which he continued to teach large numbers of influential people.

The doctrine that seems most closely associated with Pythagoras is his belief in the immortality of the soul and its repeated incarnation. His astronomical opinions are harder to define. The ideas attributed to him by the most optimistic historians may instead belong to his school. Diogenes Laertius states that Pythagoras was the first to argue that the Earth was round, although Laertius was aware that some ancient authorities held different opinions (Theophrastus gave the credit to Parmenides, while Zeno attributed this belief to Hesiod). Laertius also believed that Pythagoras was the first of the Greeks to discover that the "morning" and the "evening" star are the same (the planet Venus). Like so much else in astronomy, most historians believe that the Babylonians in fact made this discovery.

The most significant astronomical theories seem to belong to Pythagoras's followers rather than to the man himself. Aristotle, in his work *On the Heavens*, discusses the astronomical views of the Pythagorean school. The Pythagoreans had a distinct view of the universe as spherical and finite in size. They believed that the planets, the Sun, and the Moon were fixed to spheres rotating around a central fire, with the stars forming an outermost sphere. In keeping with this system, they also asserted that the Earth, Sun, and Moon were spherical in structure.

Aristotle again discusses Pythagorean astronomy in his work *Metaphysics*. He states that the Pythagoreans believed that the heavens were governed by harmony and number, arising from their studies of musical harmony. They believed that the number ten was the perfect number, and thus they believed that there were ten heavenly bodies, orbiting around the central fire. Since they could observe only nine (the first six planets, the Moon, the Sun, and the stars), they claimed that there was an undiscovered "counter-Earth" to make up this number and also to help explain the mystery of eclipses. The universe was seen to be "in complete harmony," with each of the celestial bodies being associated with its own musical note (leading to the concept of "the music of the spheres" to explain the behavior of the heavens). Because of their veneration of mathematics, the Pythagoreans believed that the distances of the planets from the central fire could be calculated through a logical arithmetical series. Thus they play an important role in our story as the first thinkers to try to use mathematical concepts to estimate the size of the universe, even though their estimates were primarily based on simplistic arguments.

The Pythagorean concept of the universe being based on spheres does seem to have been derived from detailed observations. Greek astronomers observed that in Athens the constellation Ursa Major (Great Bear) always remains above the horizon, whereas in Egypt it appears to move below the horizon. To explain these observations, they inferred that the Earth must be a spherical body floating in the sky. From this belief they drew the further inference that the fundamental shape of the celestial bodies and the heavens themselves must be spherical.

The most famous theory associated with Pythagoras is, of course, his explanation of the relationship between the hypotenuse and the other two sides of a right-angled triangle: "the square of the hypotenuse is equal to the sum of the squares of the other two sides." Sadly this theory seems more likely to have been discovered by Pythagoras's followers than by the great man himself. However, it is typical of the many discoveries and theories that the school produced in the field of mathematics, especially in the analysis of triangles and polygons.

Pythagoras certainly resembles a cult leader more than a modern scientist. The mathematical discoveries achieved by his school were, however, of lasting significance. Pythagoras deserves his reputation as the first Greek to study mathematics enthusiastically for its own sake, even if his motivation appears to have been grounded partly in religious mysticism. Pythagoras and his followers have certainly been proved right in their belief that mathematics is the essential tool to understanding the universe. For this insight alone, they do mark a significant step in the history of the understanding of the size and nature of the universe.

Anaxagoras provides the link between the cosmology of the Milesians and the philosophical schools of Athens. He was born in Clazomenae around 500 B.C.E., shortly before the Ionian revolt against the Persians. In his early twenties (according to Diogenes Laertius), he migrated to Athens, with its new democratic constitution; Athens was the largest and richest city in the Greek world at the time, after its victories against Persia. Anaxagoras was said to have written only one book, which, following the example of Anaximander, boldly attempted to give a complete account of the origin and structure of the cosmos, including some inquiry into its size. (Diogenes Laertius says that Anaxagoras was the first natural philosopher to publish a book with diagrams.)

Anaxagoras was the first philosopher to live and work in Athens—but he was the first of many. He became a close friend of Pericles, the greatest statesman of democratic Athens, and also of the radical Athenian playwright Euripides. He taught the famous moral philosopher Socrates, who eventually chose to explore very different philosophical questions. The comic poet Aristophanes parodies Socrates as an impious stargazer in his famous surviving comedy The Clouds. This picture seems very remote from the Socrates recorded in detail by such writers as Plato and Xenophon, and may in fact be a parody of Anaxagoras and his followers. Of all the philosophers we have so far discussed, Anaxagoras was the one most committed to cosmology. When someone asked him what was the object of being born, he replied: "to investigate the Sun, Moon, and heavens." According to Diogenes Laertius, when someone asked him why he was unconcerned with politics and the welfare of his country, he replied: "be quiet—I have the greatest care for my country"—pointing to the heavens.

Anaxagoras's greatest claim to fame rests on his solution of one of the great questions of ancient Greek astronomy, which dated back to Thales of Miletus, namely a persuasive explanation of solar and lunar eclipses. He discovered that the Moon does not shine by its own light but receives its light from the Sun. From this premise, he correctly deduced that eclipses of the Sun were caused by the interposition of the Moon, and that eclipses of the Moon were caused by the interposition of the Earth. This quality of reasoning shows the progress that had been made over the first hundred years of Greek cosmology and marks Anaxagoras as a rational scientist. However, Anaxagoras also agreed with Anaximenes that it was reasonable to assume that other dark heavenly bodies existed which, although invisible to us, sometimes obscured the Moon to cause eclipses.

Anaxagoras clearly also developed detailed theories to explain the origin of the universe. He believed in a "motive principle," which may be loosely interpreted as "mind." This motive principle disturbed a vortex of collected matter, where "all things were together." The rotary movement of this vortex began at a single point and progressively spread through wider and wider circles. This accelerating movement was supposed to have caused two distinct masses to separate off. The first, which Anaxagoras called "aether" or "fire," consisted of the hot, the light, and the dry, while the second, which he called air, consisted of the cool, the heavy, and the wet.

Anaxagoras assigned aether/fire to an outer circle of the universe, while air made up the inner area. The next phase of his hypothesis was the separation of the inner air into clouds, water, Earth, and other solid and liquid objects, as opposed to the gaseous substances of aether/fire. Thus Anaxagoras theorized that the heaviest mass had collected in the center, and from this mass Earth was formed. In order to explain the stars, Anaxagoras assumed that because of the violent whirling motion of the inner and outer circles, the surrounding fiery aether had torn stones from the Earth and created the stars. As ridiculous as these notions appear today, they were imaginative hypotheses to explain the observed natural world.

Anaxagoras also had detailed theories about the observable heavenly bodies. Following Anaximenes, but disagreeing with the Pythagoreans, he believed that the world was flat and was supported by air. Anaxagoras claimed that the Sun and stars were all stones on fire, carried round by the revolution of the aether. He did attempt to estimate the size of some of the heavenly bodies, arguing that "the Sun is larger than the Peloponnese" (the peninsula of southern Greece). This does seem to imply that he did not think it was much bigger than the Peloponnese, and particularly that he thought that the Sun was smaller than the Earth. However, we do not know Anaxagoras's reasoning to produce this belief. He is also quoted as believing that the Moon was of a substance similar to the Earth, and that it had plains and mountains (a fact that Galileo would verify when he first turned a telescope to the heavens in 1609 C.E.).

For the quality of his reasoning, the depth of his imagination, and above all for his absolute commitment to scientific reasoning against the opposition of powerful political persecutors, Anaxagoras undoubtedly deserves a place among the most significant of the Greek thinkers. There is good evidence that Anaxagoras suffered persecution from the institutions of democratic Athens for his scientific beliefs. Plutarch, writing under the Roman Empire, records that the "decree of Diopethes," the diviner, was directed against Anaxagoras. This decree, presumably passed by the assembly of Athenian citizens, ordered that "anyone who did not believe in the gods or who taught theories about celestial phenomena" should be liable to prosecution—presaging a form of intolerance to intellectual advancement witnessed again when Copernicus and Galileo sparked their own revolution in cosmology. Anaxagoras was accused of impiety, allegedly for saying that the Sun was a red-hot stone and the Moon made from Earth. This sad affair shows that even in the fifth century B.C.E., at the height of its prosperity, the Athenian democracy was capable of the kind of witch hunt practiced again in 399 B.C.E. with the trial and execution of Socrates, five years after the demoralizing defeat in the Peloponnesian War against Sparta.

After Anaxagoras, the quantity of information on the ancients increases dramatically. For all the previous philosophers, we have had to rely on fragments or short accounts of their work. By contrast the surviving works of the great classical philosophers Plato and Aristotle run into several volumes.

We will next move on to consider the view of the cosmos produced by the scholars of Plato's Academy, an institution where educated scholars and mature students researched the deepest questions of classical Athens. Plato's own philosophical interests developed from the questions examined by his great teacher Socrates. Thus Plato's great works are mainly concerned with ethics, politics, epistemology (the study of what we can know), and the immortality of the soul. He believed that reality could be understood only through gaining knowledge of ideas or forms, such as the form of the good, and that the objects of the visible world were at best inadequate copies of the real forms, which had existed since the beginning of time. Thus Plato and his subsequent followers tended to reject an observational and experimental approach to astronomy, because of distrust of the senses and because of their belief in rational arguments from first principles.

Despite this idealistic epistemology, Plato did write one major dialogue, the *Timaeus*, where he outlines a series of cosmological beliefs. Plato's own position on the explanation for the creation of the universe comes down clearly on the side of the argument from design. The most powerful narrative within the *Timaeus* is that of the great designer, the demiurge, creating the cosmos as a living being according to a perfect model. A deity imposes reason on necessity to bring order from the receptacle of disordered matter, creating a "child," the cosmos, which is the copy of a perfect idea, which has existed eternally. Plato adapted Anaxagoras's belief in "mind" controlling the universe and supposed that a specific intelligence had organized the creation of the universe and governed the laws of nature. Here we have a total commitment to an ordered and hierarchical cosmos—with no hint of its opposite, chaos.

In his classic work the *Republic* Plato seems to assume an early Pythagorean theory of the solar system, with the heavenly bodies rotating round the Earth. Plutarch and other writers refer to a later Platonic view, that the Earth was not worthy of the central place in the universe, and these writers say that he adopted the later Pythagorean theory of the central fire around which the planets rotated. This view gains some support from a passage from the *Laws*, Plato's final work, written in his extreme old age. The Athenian stranger, who most commentators think represents Plato himself, says that the Sun, Moon, and planets "always follow the same path in a circle."

Despite this wealth of material, Plato seems to have popularized Pythagorean views of astronomy rather than to have developed original ideas (in sharp contrast to his contribution to epistemology and political thought). His main contributions were to follow Pythagoras in stressing the value of mathematics and to create a school where scholars could think and study. It was later members of the Academy who would make the school's most significant contribution to astronomy.

As well as being a great philosopher, Plato was undoubtedly a great educator, who particularly believed in the educational value of astronomy. He is said to have set as a problem for his successors at the Academy the challenge to find "what are the uniform and ordered movements by the assumption of which the motion of the planets can be accounted for?" Plato was certain that circular motion was the key, since this was the simplest uniform motion that repeated itself endlessly as the annual cycle of the heavens appeared to do.

Eudoxus of Cnidos produced the first development in astronomical theory in response to this challenge. Eudoxus had attended some of Plato's lectures at the Academy and was a talented mathematician. Eudoxus's approach to explaining the motion of the heavens was the hypothesis of concentric spheres. Eudoxus assigned the fixed stars to a huge outer sphere and the Earth to a much smaller sphere fixed at the center. The huge star sphere then rotated around the Earth every twenty-four hours. The Sun was attached to a concentric sphere within the star sphere, and clearly its sphere had to be transparent since the stars could be seen through it at night. Then more concentric spheres were assigned to the planets. However, this "simple" arrangement could not describe the perceived motions. A particular issue was the behavior of some of the planets. What made them pause their easterly motion night on night and, for a short period, appear to trace a westerly loop before resuming their eastward path—a so-called retrograde motion? Eudoxus reproduced the irregular planetary motions through adding further concentric spheres to the planetary concentric spheres, each revolving at a uniform rate but about different axes. He needed four extra concentric spheres for each of the planets. And he needed to allocate three extra spheres to the Sun and the Moon to describe their motions. By adjusting the orientation of each axis, and the rotational velocity of each and every sphere, Eudoxus was able to reproduce, with a reasonable approximation, the motions of the celestial bodies as they had been observed. Eudoxus's spheres were subsequently modified by a later astronomer, Callipus, and then adopted and modified by Aristotle, who added more spheres to the celestial bodies. Aristotle's synthesis of "prior art" in fact contained fifty-five transparent spheres! Aristotle clearly believed in the reality of these transparent spheres, although it is likely that Eudoxus saw them merely as a clever geometrical trick to predict the behavior of the planetary bodies. Eudoxus's hypothesis was the first detailed attempt to provide a mathematical basis to explain the observations of the solar system. It was an elegant (albeit incorrect) use of geometry applied to an astronomical problem, based on the established Pythagorean tradition of the perfection of the sphere.

Before Eudoxus, Greek astronomers were vague about the number and names of the planets. Eudoxus's hypothesis and calculations ended this ambiguity and imposed greater discipline and accuracy on astronomical studies. Eudoxus seems to have been well aware of Babylonian discoveries. In his work he does make use of Babylonian observations, which can be seen in his descriptions of different constellations. Much of what followed owed immeasurably to the foundations laid by Eudoxus.

Heraclides of Pontus (388–315 B.C.E.), who was a pupil of Plato and a contemporary of Aristotle, made further important contributions to astronomy. Heraclides was the first Greek astronomer to argue that the apparent daily rotation of the heavenly bodies is caused not by a rotation of a heavenly sphere about an axis through the center of the Earth but by the Earth itself rotating about its own axis. Heraclides' second argument was that Mercury and Venus orbit around the Sun. These two theories anticipate the Copernican revolution, and another Greek astronomer, Aristarchus of Samos (310–230 B.C.E.), as we shall see, further anticipated Copernicus's theories by arguing that all the planets (including the Earth) rotate around the Sun. Sadly these major advances in understanding were not consolidated into later developments, and the logic of the planets orbiting the Sun was lost.

Alongside Plato, Aristotle is the other genius of Greek philosophy whose work survives in extensive detail. Although Aristotle's work does not have the outstanding literary merit of Plato's dialogues, his arguments in fields such as ethics, logic, politics, and natural science have probably made an even greater contribution to our civilization.

Aristotle (384–322 B.C.E.) was born in the northern Greek kingdom of Macedonia, in the town of Stageira. In his younger days he became one of the most famous students and scholars of Plato's Academy, but around 344 B.C.E. he founded the Lyceum, which became the second of the great philosophical schools in Athens. His followers were called Peripatetics, after the gathering place called Peripatos (the Walk), which was in the same area of Athens. An important motivation for Aristotle to set up the Lyceum was the fact that he had been overlooked for the post of head of Academy on Plato's death, which went instead to Plato's cousin, Speusippos (who in fairness was also an important thinker).

Like Plato, Aristotle believed that astronomy was a vital part of science and that its study was of great educational value. Also like Plato, however, he did not produce much original work on astronomy. Aristotle's major contribution to the history of cosmology was to provide testing criticism of the work of earlier natural philosophers, and it is through such criticism, and from that of later commentators on his work, principally Simplicius in the sixth century C.E., that we get much of our knowledge and understanding of earlier cosmological writers.

Aristotle's analysis of the different cosmological theories was governed by his differences from Plato in his views on the acquisition of knowledge. Aristotle was skeptical of Plato's theory of perfect and eternal forms, which might potentially be understood by the use of reason alone. Instead he believed that observable phenomena made up the real world, whereas ideas and concepts explained the essence of these observable phenomena.

Aristotle followed Empedocles, the Greek philosopher from

Sicily, in believing in four basic elements: earth, air, fire, and water. However, Aristotle also believed in a fifth element (the "quintessence"), which he called "aether." All the heavenly bodies, he surmised, were made from this fifth element. While the four terrestrial elements continued to change from one to another, the element belonging to the heavens remained unchanged and perfect. Thus was born the philosophy of the perfect immutable universe—and this Aristotelian view of the cosmos would remain largely intact in Europe until the Renaissance.

Aristotle accepted the vision of the cosmos developed in the model of the spheres of Eudoxus. Indeed it is thanks to Aristotle's work that we have detailed records of Eudoxus's hypotheses. Aristotle also presented a number of important astronomical arguments in his own right. Firstly he presented some logical inferences from observation to prove that the Earth must be spherical. He noticed that the shadow of the Earth always appears as an arc on the Moon during a lunar eclipse. He also confirmed that the star sphere appeared displaced as an observer moved north or south on Earth's surface, that eclipses occurred at different times at different locations, and that in sailing toward a distant island it appeared to emerge up out of the sea. All these observations, he argued, indicated that the Earth had to be spherical.

Aristotle believed that the Earth was at the center of the universe and was farther from the Sun than from the Moon. His reason for this second belief was that during a total solar eclipse the Moon completely covered the Sun. Of course this deduction is correct, and is in keeping with the quality of Aristotle's best logical deductions.

Aristotle believed that the universe was spherical in structure and had a constant circular motion. Thus he deduced that it must be finite, since an infinite body has no center around which it may rotate. He believed there could be no space outside the universe, since space is only that in which a physical body can exist.

As we have already discussed, Aristotle was influenced by Eudoxus's hypothesis of concentric spheres. The fact that Aristotle added further "reacting spheres" to try to produce a coherent system to govern the movement of all the heavenly bodies made the hypothesis more complicated and represented no significant improvement in the ability to track the paths of the Sun, Moon, and planets.

After the conquests and death of Alexander in 323 B.C.E., the center of Greek astronomical research gradually moved from Athens to Alexandria in Egypt. As Alexander's empire was breaking up, Ptolemy, one of Alexander's friends and marshals, seized control of Egypt and set up a Greek kingdom there. Ptolemy's Egypt turned out to be the longest lasting of the Greek monarchies founded by Alexander's successors, and the famous Cleopatra was the last of the Ptolemaic dynasty, losing her kingdom to the Roman leader Octavian (later Augustus) in 30 B.C.E.

The Ptolemaic dynasty enters our story because they were the greatest patrons of science in the Greek world. Ptolemy had plenty of money from his rich revenues and his wealthy capital, Alexandria. However, he wanted legitimacy and social standing for his new kingdom without provoking stronger military powers such as the Seleucids. What better way to achieve this than to sponsor the Greek world's leading scientists? As in the modern world, the academics quickly followed the money, resulting in an ancient brain drain. Ptolemy's successors continued his enlightened policy, resulting in a golden age of patronage. Thus they set up the final great Greek astronomical school, called the Alexandrian Museum, which sponsored great scholars, such as Aristarchus, Eratosthenes, and Hipparchus. In fact the most significant developments in Greek astronomy were made at Alexandria in the second century B.C.E. Alexandrian astronomers were also helped by easier access to the Babylonian data and by greater exchange of information all around the Mediterranean and the Near East, which was under control of Greek kingdoms. The library of Alexandria became the greatest archive of classical learning until its tragic destruction in the seventh century C.E.

The first great astronomer of the Alexandrian Museum was Aristarchus (about 300–230 B.C.E.). Aristarchus was born on the Greek island of Samos, the birthplace of Pythagoras, but migrated to Alexandria—presumably to benefit from the patronage of the Ptolemys. Aristarchus's great claim to fame was that he was the first

Greek astronomer to argue on the basis of logic that the Earth orbits the Sun. Aristarchus's theory was that the Sun and the stars are fixed bodies. He placed the Sun at the center of the universe, while the stars were located on a distant outer sphere. Earth and the other planets revolved around the Sun in circular orbits. Aristarchus's contemporaries and successors, such as Hipparchus, the greatest of all the Alexandrian astronomers, nearly all rejected his theory. His only supporter, says Plutarch, was Seleucus of Seleucia on the Tigris. Archimedes describes the Greek scholars' objection as being that there was no observable change in the apparent position of the stars, such as one would expect if the Earth moved around the Sun. (This is the parallax issue, which we will explore in some detail in the next chapter.) Since their current theories appeared to explain the observable phenomena, these astronomers, who founded their views on Aristotle's epistemology, had no incentive to change these theories.

Aristarchus did produce the correct riposte to this objection, namely that the stars were at such a distance from Earth that it was impossible to observe any apparent motion in the stars. Thus, for our story of the different theories on the size of the universe, it is interesting to note that of all the ancient astronomers, Aristarchus is the one who conceived of the vast distances from Earth to the stars. However, he could produce no evidence to prove this assumption, and the consensus of opinion swung away from a heliocentric theory. Aristarchus produced an improved estimate of the distance from the Earth to the Moon from simple geometrical arguments, deriving a pretty fair estimate of thirty Earth diameters. This Earth-Moon distance would provide a reasonably secure first rung on the cosmic distance ladder for almost two thousand years.

Aristarchus's theory of a heliocentric solar system with the stars a vast distance away, which he proposed in the third century B.C.E., was remarkably accurate, but it was sadly ignored by scholars until the theories of Copernicus in the fifteenth century C.E. In fact, like Anaxagoras, Aristarchus attracted religious criticism. Plutarch records that Cleanthes the Stoic thought that Aristarchus ought to be indicted for impiety "for putting in motion the Hearth of the Universe." Happily Ptolemy's kingdom of Alexandria was more tolerant of the conclusions of astronomers than democratic Athens had been—or the medieval Inquisition would prove to be.

Eratosthenes, the third head of the Alexandrian Museum, was the first man to be able to accurately estimate the size of the Earth, and he achieved this by astronomical means. By estimating the size of the Earth, Eratosthenes also achieved an important step in estimating the size of the cosmos.

Eratosthenes' method for measuring Earth's size was a brilliant combination of logic and geometry, as recorded by the writer Cleomedes. On June 21 (the first day of the summer solstice) he observed in the town of Syene (modern Aswan) in the south of Ptolemy's kingdom that the noon Sun was directly overhead, because its rays completely covered the floor of a deep well. However, on the same day in Alexandria, a distance 5,000 stadia north, the Sun was not directly overhead, because objects there were casting shadows. He assumed that the Sun's rays reach Earth along parallel lines. He then measured the angle between the Sun's rays and an obelisk in Alexandria, which he estimated at 7½ degrees. He then deduced that this angle was equal to the angle at Earth's center that separates the diameter to Alexandria and the diameter to Syene. Since 7½ degrees is ⅟₄₈th of the 360 degrees making up a complete circle, he deduced that the distance between Syene and Alexandria must be ⅟₄₈th of Earth's circumference. Thus he multiplied the distance between the cities by 48 and produced a value for Earth's circumference (within an admirable precision). The quality of this deduction, the use of observation, and the application of geometry make it one of the most striking achievements of the school of the Alexandrian Museum in its long and illustrious history.

Most scholars consider Hipparchus to be the greatest observational astronomer of the ancient Greek world. He was born in Nicaea, in Bithynia (modern northwestern Turkey), but, like so many other Greek scientists, migrated to Alexandria. Claudius Ptolemy records his observations between 161 and 126 B.C.E., so we can deduce that he was born around 185 B.C.E. Hipparchus's unique achievement was the creation of a catalogue of nearly one thousand stars. He created this catalogue with his own detailed observations and by carefully examining the observations of his pre-decessors. He had the advantage by the latter part of his career of 150 years of records built up by the Alexandrian Museum, and he also used data of Babylonian origin. Hipparchus's catalogue was so accurate that it enabled later astronomers to make important discoveries.

Hipparchus was also interested in calculating the movements of the Sun, Moon, and planets. From detailed observation of the position of the stars he was able to demonstrate that Earth's axis of rotation is not fixed in space but precesses gradually. He calculated the length of seasons and developed a chart that gave the position of the Sun on the ecliptic for each day of the year. He made an excellent estimate of the tropic year as 365 days, 5 hours, 55 minutes, and 12 seconds. (This exceeds the modern calculation by a mere 6 minutes and 30 seconds.) Certainly Babylonian calculations by astronomers such as Naburiannu (fl. 500 B.C.E.) and Kidinnu (fl. 383 B.C.E.) would have helped Hipparchus significantly. Under Hipparchus precision observational astronomy was coming of age.

In order to produce his observations, Hipparchus is credited with developing a number of measuring devices. He is believed to have used the armillary astrolabe, a set of concentric rings rotating round one another to simulate the relative movements of the heavenly bodies. He also used an improved dioptra (a primitive surveying instrument for measuring angles), which could be adjusted for the inclination of the North Pole.

Significantly Hipparchus introduced the idea of an "epicycle," a small circular orbit rotating around a big circular motion, as a mathematical alternative to concentric spheres in an Earth-centered cosmos, a concept extended later by Claudius Ptolemy. But to make use of the concept a form of trigonometry was needed, which Hipparchus developed. Thus Hipparchus can be credited with the invention of trigonometry. He compiled a table of chords in a circle, an early version of modern trigonometric tables. Together with the work of his catalogue of the stars, Hipparchus also devised a magnitude scale for measuring the brightness of stars, a derivative of which is still used today.

Importantly for our story, Hipparchus improved on Aristarchus's estimates for the distances of the Sun and Moon. He calculated the mean distance of the Moon to be about 34 times the diameter of the Earth and the distance of the Sun to be 1,245 times the diameter of the Earth. This estimate for the Moon is remarkably accurate, but he underestimated the distance to the Sun by a significant margin. These calculations show that Hipparchus had gained some understanding of the vast size of the solar system. However, it would take more modern techniques to understand the vast distances to the stars.

Hipparchus's career is a clear contradiction of the generalization that Greek thinkers were concerned mainly with theory to the exclusion of experiment and measurement. This misunderstanding has developed from an analysis that has placed too much weight on Plato's epistemology and not enough on the works of Aristotle and his successors. Hipparchus's work is so remarkable precisely because it is based on detailed observation, accurate measurement, and sound logical reasoning.

The Alexandrian Museum and library continued to be proud beacons of Hellenistic culture, but power and wealth were moving west. The developing power of Rome threatened the stability of the empire, and while Alexandria continued to be a populous and wealthy city, less original research was produced after the death of Hipparchus. Rome's culture was more militaristic, and the golden age of patronage was over.

After 30 B.C.E. Egypt and Alexandria were integrated within the Roman Empire, as a special fiefdom of the emperor. Its strategic importance became vital as Rome's major source of grain. No Roman senator was allowed to enter Egypt without the emperor's permission. Alexandria became the empire's second city and had the most turbulent civil life, with regular riots between the different ethnic populations, such as Greeks, native Egyptians, and Jews, all carefully monitored by the governor of Egypt, one of the most sensitive posts in the whole empire, who reported directly to the Roman emperor. The proud traditions of the Alexandrian Museum and the library continued, respected and honored by the Roman conquerors, but fewer great names emerge from our sources to match the eminence of Aristarchus, Eratosthenes, and Hipparchus.

The one great astronomer of Roman Alexandria was Claudius Ptolemy. He flourished in Alexandria around 150 C.E., at the height of the Roman Empire. His very name was a sign of the times, with his second name recording the Hellenistic rulers but his first name recalling an emperor of Rome.

Claudius Ptolemy's great contribution to astronomy was his famous work the *Almagest*, which presented formally the astronomical theories of the day that had evolved from the great debates within the different Greek philosophical schools. Claudius Ptolemy freely admitted that he had contributed little original research to the treatise but rather had based his conclusions principally on the work of Hipparchus, nearly three hundred years earlier. The *Almagest*, which survives in detail, is by far our most important source for the work of Hipparchus. Ptolemy, like most scholars, preferred Hipparchus's Earth-centered universe to Aristarchus's heliocentric system. And he built up the use of circular orbits with epicycles in an Earth-centered universe as the way to describe the complex motions of the heavenly bodies—including the retrograde motion of the planets. Each planet moved in a small circle, called the "epicycle," the center of which moved around a larger circle, called the "deferent." The retrograde motion of the planet coincided with the planet passing inside the deferent.

Importantly, Ptolemy did not claim that his cosmological model described the actual conditions. It simply reproduced geometrically the observed motions of the known heavenly bodies and enabled their positions to be easily predicted for any particular time. For over fourteen centuries, the *Almagest* was accepted as the prime source of knowledge on the theories of Greek astronomy and was used as the basis for astronomical work. As the religious beliefs of Europe changed, Ptolemy's work was accepted by the Catholic Church and was assumed into the canon of orthodoxy. The mind of humanity was fixed for almost fifteen hundred years. Only the work of the great figures of the scientific Renaissance would shift it. The Renaissance thinkers are of course vital in our quest, and they made a huge contribution to how educated people view the universe. Ironically, even when Copernicus's heliocentric theory had replaced the Ptolemaic system, many astronomers used

Ptolemy's model to predict the motion of the planets, since its intricate calculations produced more accurate values.

In fact modern astronomical detective work suggests that Ptolemy may have falsified certain observations to demonstrate the validity of some of his ideas. Of course he could never have anticipated that the power of modern astronomy would catch up with his sleight of hand. However, this should in no way detract from his monumental works.

From Thales to Hipparchus, the great age of innovation in Greek astronomy covers over four hundred years. The *Almagest* of Ptolemy represents the climax of their debates, and the strictures of authority restricted research in Europe beyond the conclusions of the Greeks until the heroic discoveries of Copernicus, Tycho, Kepler, Galileo, and Newton a millennium and a half later. We may ask, What distinguishes the Greeks' astronomy from that of their main competitors from the ancient world, the Babylonians? The answer is the genius of the Greek philosophers for speculation. The Greeks owed a great debt to the Babylonians in terms of the details of their astronomy, and before Hipparchus they were inferior to the Babylonians in the quality of their observations. However, the variety of speculative visions from the Greeks was unique before the Renaissance. We know that the Babylonians produced detailed cycles for the prediction of eclipses and that they observed and recorded the apparent motions of the planets; however, we have no evidence of the Babylonians formulating any theories to account for the movements of the Sun, Moon, and planets.

Over successive generations, Greek natural philosophers such as Anaximander, Anaxagoras, Aristarchus, and Hipparchus developed bold theories, containing elements that have provoked interest among all subsequent generations of cosmologists. The ingenious visions of the Greeks and their speculative genius provided inspiration for the great thinkers of the scientific renaissance, who rediscovered the Greek joy of original thinking. The four hundred years we have concentrated on, from Thales to Hipparchus, were an era of progressive development and continuous debate. The debates between the followers of Plato and Aristotle, and especially between the adherents of Hipparchus and Aristarchus, would estab-

lish a tradition of confrontational dialogue between different philosophies that has characterized scientific advancement up to the
present day.

The Greek passion for knowledge and discovery should certainly
be a challenge to our own era, in which the idea of astronomical
research, and indeed academic study for its own sake, is under constant attack. A fitting epigram to summarize the spirit of the Greek
natural philosophers was written by the last of the great line, Claudius Ptolemy.

I know that I am mortal and the creature of a day; but when
I search out the massed wheeling circles of the stars, my feet
no longer touch the Earth, but side by side with Zeus himself, I take my fill of ambrosia, the food of the gods.

Chapter 2

SERIOUS MEASUREMENTS

The teachings of the ancient Greeks dominated intellectual development within Europe from Claudius Ptolemy all the way through to the sixteenth century. But for a long while there was little further development. With the Mongols and the Chinese pressing them from the east, the Huns started a westward path of conquest in the third century C.E. They destroyed much of what they encountered. The collapse of the great Roman Empire was finally signified with the deposing in 476 C.E. of the last emperor of the West, Romulus Augustulus, by Odoacer. The successor Byzantine Empire followed suit when the Turks captured Constantinople in 1453 C.E. From 400 C.E. scholarship, so cherished by the Greeks, declined in Europe because of the hostility between the Christian church and paganism. The great schools of Greece, and the Alexandrian Museum, had been pagan. The Christians destroyed many of the institutions that were perceived as being pagan and burned many of the great classical libraries such as that at Serapis. Much of the precious cultural heritage of the ancient Greeks was destroyed in burnings of the books in the name of Christian orthodoxy. This destruction of ancient writings was a crime of immense proportions. The Western world entered an intellectual dark age. During this period it was the Arabs who assumed custody of the proud Greek heritage of scholarship.

The *Almagest* was translated into Arabic and formed the basis for a new golden era of Arabic astronomy. The Arabs proved to be skilled observers, and they established centers in Baghdad and Damascus to advance astronomy. Novel equipment was built for observing the stars. Al Mamon, one of the finest scholars of the ninth century, built an observatory and astronomical library in Baghdad that was the finest of the era. The best-known Arabian astronomer of the ninth century was Al-Battani. The quality of his observations has been favorably compared with those of Hipparchus. Arab scholars advanced mathematics in particular, introducing algebra as an alternative to Greek geometry to solve scientific problems, and some of their advances reentered Europe through Spain from about the tenth century. As far as astronomy was concerned, European intellectuals were unable to offer anything new for over a thousand years. However, the complexity of the system inherited from Ptolemy, with its system of cycles and epicycles, did not escape critical comment. While watching his court astronomers struggle with predicting the motions of the planets in the middle of the fifteenth century, King Alphonso X of Castile complained that if he had been present at the creation, he would have suggested to the Almighty a somewhat simpler arrangement. (Alphonso was responsible for getting Arab and Jewish scholars, whom he had summoned to Toledo, to produce the famous *Alphonsine Tables* for forecasting eclipses. These tables were used for almost three hundred years.)

The end of the Arabian dominance of astronomy came in a bizarre fashion. Ulugh Beg was grandson of the Asian conqueror Tamerlane. His father had captured the city of Samarkand and gave it to Ulugh Beg. He turned it into a city of Muslim culture and constructed a magnificent observatory there in 1428 C.E. In fact his principal interest was astrology, rather than the furtherance of science. He produced a horoscope that predicted that his eldest son, Abd al Latif, would kill him. In an attempt to escape this destiny, he banished his son from the kingdom. The disgruntled son initiated a rebellion against his father, whom he ordered to be killed (thus fulfilling the father's prognostication) . As the successor ruler, Abd al Latif destroyed much of the cultural infrastructure his father had put in place, which he perceived as having been used against him. Not surprisingly, later rulers were not as attracted by astrology as had been Ulugh Beg. The era of Iranian cultural dominance was ended.

There were five individuals who during the sixteenth and seventeenth centuries finally broke the Ptolemaic stranglehold inhibiting intellectual advancement and took scholarship in Europe forward from the levels achieved by the ancient Greek scholars. They were Nicolaus Copernicus, Tycho Brahe, Johannes Kepler, Galileo Galilei, and Isaac Newton. There were many other creative thinkers during a century and a half of scientific awakening from the mid–sixteenth century until the start of the eighteenth century, but these five did more than any others to redefine human understanding of the cosmos. Their collective contributions must be deemed to be equivalent to the monumental contributions made by the Greek schools in their day. Together the five Renaissance scholars established a new epoch in understanding and redefined the capabilities of the human intellect. Other than Tycho and Kepler, who collaborated only briefly, they all worked separately. But the five were united by their individual belief that an understanding of the creation was within the grasp of scientific reason.

The epoch of new enlightenment was founded on the thinking of Nicolaus Copernicus, who was born in 1473 C.E. in Toruri, Ermland (then under the Polish crown). From 1491 until 1494 he studied mathematics and classics at the University of Krakow. In 1496 he traveled to Bologna, Italy, to study law and later astronomy. He studied both secular and ecclesiastical law. His uncle and patron, Bishop Watzenrode, was able to get him appointed canon of Frombork, a post he retained notionally for the rest of his life. The post carried only light duties (which was fortunate, since Copernicus was distracted by study and family duties) but provided an adequate income. Thus Copernicus was able to devote much of his time to scholarship in general and astronomy in particular without having to face financial worries. In 1500, after completing his studies at Bologna, Copernicus moved on to Padua to study law and Greek. He gained a doctorate in canon law in 1503 in Ferrara and then returned to Padua to study medicine. By any measure the extent of his learning was impressive.

With his mastery of Greek, Copernicus turned his attention to reading practically all the works of the great Greek astronomers. His inquiring mind could not accept all that he read as a divine insight, and he realized that despite the intrinsic beauty of many of the arguments of the ancients, they mainly defied philosophical logic. Perhaps it was the diversity of his learning that made it possible for him to think laterally and challenge conventional wisdom. Or perhaps he found particular fascination in what he had read about the ideas of Aristarchus, suggesting a universe centered on the Sun around which the planets and stars orbited.

In 1506 Copernicus returned home to Poland to serve his uncle as personal doctor and secretary. In about 1513 he wrote a brief outline of some new ideas on astronomy he had developed, challenging the Earth-centered model for the universe of Claudius Ptolemy as being overly complex, unwieldy, insufficiently accurate, and counter to intuitive logic. Copernicus felt that all the ancient concepts of the universe based on the complex interplay of spheres and circles to explain the motions of the celestial bodies were wrong — both scientifically and philosophically. Copernicus's alternative model placed the Sun at the center of the solar system (which lay near the center of the universe), with Earth and other planets orbiting around it. Although Aristarchus had postulated just this concept some eighteen hundred years earlier, such a heliocentric vision of the universe had been buried under Aristotelian and Ptolemaic dogma, and the voices of reason had been drowned out by the cacophony of the music of the spheres.

The basic elements of the Copernican model were straightforward. He had Earth rotating daily about its axis. Ptolemy had rejected this idea; he argued that a rotating Earth would fly apart and destroy itself. But Copernicus exposed the nonsense of this notion, pointing out that on the same basis Ptolemy's more massive rotating celestial sphere would have destroyed itself long since. Copernicus had all the celestial bodies orbiting the Sun such that the closer to the Sun a planet was, then the greater its orbital velocity.

The beauty of the Copernican model was the way it handled the retrograde motion of the planets. Retrograde motion, you will recall, is the apparently periodic looping of the planets in their movement across the heavens, their normal easterly passage against the background stars being interrupted by a westerly loop before the easterly drift is resumed. The Greeks had been forced to give the celestial bodies epicycles, miniature orbits superimposed on their main orbits, to describe the retrograde behavior—a geometrical trick that no one could give any plausible physical explanation to. And to give Ptolemy his due, he saw epicycles as a tool for prediction rather than a description of reality. But in the Copernican model, with Earth and other planets orbiting the Sun and their orbital velocities increasing with proximity to the Sun, the retrograde motion had a simple explanation. Consider Earth orbiting the Sun in one year. The next planet out from the Sun, Mars, takes about two Earth years to orbit the Sun. Thus as they each orbit the Sun at different rates Earth periodically overtakes Mars, which appears to briefly remain stationary. As Earth passes Mars it appears to move in a retrograde motion, and then as Earth moves away Mars again appears to move in its easterly path. Thus by assuming that the Sun is at the center of the universe with the planets all revolving around it at different rates, Copernicus was able to explain in a simple and natural way the retrograde motion of the planets.

Copernicus certainly did not get everything right; not least he insisted on circular orbits, because of his continued belief in the perfection of a circle. Hence in his heliocentric model planetary motions could not yet be predicted with certainty.

Copernicus privately circulated a comprehensive description of his model in 1530, including it first in a letter to a friend. This letter was copied and widely circulated, with Copernicus's agreement, and became known as the *Commentarioulus*. Realizing that he risked being branded a heretic because of his anti-Ptolemaic views, Copernicus was reluctant to publish his ideas too widely. His reasoning with ecclesiastical colleagues was that he was promoting his model as an interesting and intuitive idea but was not wishing to challenge church doctrine. He was finally persuaded by close friends to publish his theories in a book, *De Revolutionibus Orbium Coelestium*, which he prudently dedicated to Pope Paul III in the hope that this would allay fears of the church that he was seeking to challenge its teaching. As an extra safeguard his publisher inserted a disclaimer saying that the theory was merely a mathematical fiction. Copernicus realized that the absence of a measurable parallax required the stars to be at a vast distance. He still envisaged the stars as all being at the same distance, that of the celestial dome, but gave recognition to their extreme remoteness: "So vaste, without any question, is the divine work of the almighty Creator."

Copernicus received a copy of the delayed printed work only on his deathbed. He thus escaped the revolution in scientific and theological thinking his work was to precipitate following his death in 1543. While Copernicus's death meant he avoided being branded a heretic, others who promoted his cause were not so lucky. And the church eventually decided to put his *De Revolutionibus* on the index of banned books since some people were putting far too definite an interpretation on what the church thought was merely a hypothetical description.

Copernicus's work was not easy to understand, especially by scholars steeped in Ptolemaic teachings and church dogma. But there was one person who understood its true significance and was committed to promoting the truth. He was Giordano Bruno, a philosopher of distinction and a strong supporter of Copernicus. He took upon himself the task of promoting Copernican thinking, through lectures and writing. Not only did Bruno promote the Copernican Sun-centered universe; he also presented the case for an infinite universe. He argued that the planets were seen in reflected light from the Sun, while the stars were very hot bodies radiating their own light just as the Sun did. But, he suggested, the stars were at vast distances and might have their own system of planets, on which creatures, perhaps more or less intelligent than humans, might live. Although the theories of Bruno represented prophetic insight of the most profound form, his teachings challenged the primacy of church doctrine. The Inquisition insisted he recant such heretical views. He would not. In 1600, after seven years in prison during which he still refused to recant, the Inquisition had him burned at the stake in the Campo del Fiori in Rome. His final words were: "The time will come where all will see what I have seen."

Three years after the death of Copernicus, the great Danish astronomer Tycho Brahe was born. He was of noble birth. He had an uncle of exceptional wealth, and in a strange pact Tycho's parents had promised the childless uncle that he could bring up their firstborn son and give him the very best education money could buy. The uncle always sought the very best for the young Tycho in all things. He sent him to the university of Copenhagen to study law and languages, the basic ingredients of statecraft, which the uncle hoped Tycho would practice.

While at the university, a predicted solar eclipse visible from Copenhagen fascinated the young Tycho, and he turned his hand to mathematics to try to understand how the eclipse had been predicted. An intense interest in astronomy followed naturally from this first experience of studying a cosmic phenomenon. He acquired the *Almagest* and mastered its intricacies. His uncle was less than pleased with this new interest and sent his nephew to Leipzig with a private tutor instructed to keep him focused on his law and language studies, and not distracted by astronomy. The tutor failed in this regard, but Tycho kept his growing interest in astronomy secret from his uncle. Following his uncle's death, and a substantial inheritance, Tycho Brahe was able to dedicate the rest of his life to astronomy.

After his period of study at Leipzig, Tycho went to the university of Rostock to further his learning in astronomy. While there he got into a heated argument with another scholar (the topic of disagreement was never revealed), and a duel with swords followed. Tycho came off the worse, losing the tip of his nose. A gold, silver, and wax replacement was designed to make good the loss.

Tycho built himself a splendid observatory at Augsburg in Germany, equipped with the finest measuring instruments money could buy. Of course this was the pretelescopic era, so measurements of the heavens were limited to what one could discern with the naked eye. He built a magnificent quadrant to measure the angle of elevation of any star and a sextant to measure angular distances between them. Tycho started measuring the positions of the stars as well as the motions of the planets. And he also discovered a majestic new star of exceeding brilliance in the constellation Cassiopeia. (Successor astronomers would recognize this event to have been a stellar explosion, of the type we now call a supernova—one of a few ever discovered in the Milky Way. None has been discovered within the Milky Way since the advent of the telescope for astronomical purposes in 1609.) Tycho recorded his amazement at finding the new star:

When on the above mentioned day [November 11, 1572], a little before dinner, I was returning to that house, and during my walk contemplating the sky here and there since the clearer sky seemed to be just what could be wished for in order to continue observations after dinner, behold, directly overhead, a certain strange star was suddenly seen, flashing its light with a radiant gleam, and it struck my eyes. Amazed, and as if astonished and stupefied, I stood still, gazing for a certain length of time with my eyes fixed intently upon it and noticing that same star placed close to the stars which antiquity attributed to Cassiopeia. When I had satisfied myself that no star of that kind had ever shone forth before, I was led into such perplexity by the unbelievability of the thing that I began to doubt the faith of my own eyes, and so, turning to the servants who were accompanying me, I asked them whether they too could see a certain extremely bright star when I pointed out the place directly overhead. They immediately replied with one voice that they saw it completely and that it was extremely bright. But despite their affirmation, still being doubtful on account of the novelty of the thing, I enquired of some country people who by chance were travelling past in carriages whether they could see a certain star in the height. Indeed, these people shouted out that they saw that huge star, which had never been noticed so high up. And at length, having confirmed that my vision was not deceiving me, but in fact that an unusual star existed there, beyond all type, and marvelling that the sky had brought forth a certain new phenomenon to be compared with the other stars, immediately I got ready my instrument. I began to measure its situation and distance from the neighbouring stars of Cassiopeia, and to notice extremely diligently those things which were visible to the eye concerning its apparent size, form, color, and other aspects.

Tycho's amazement is hardly surprising, bearing in mind the strength of the church's teaching of the Aristotelian vision of a perfect and nonchanging universe. New stars were simply not expected to appear. But the quotation also emphasizes the diligence of Tycho, who, having convinced himself of the reality of the new star, set about measuring it with great care.

Although still only a young man, Tycho quickly established himself as an astronomer of remarkable talent. The Danish monarch King Frederick II was so impressed with the young man that he invited Tycho to take the position of professor of mathematics and astronomy at the University of Copenhagen and subsequently to become the court mathematician. The king presented Tycho with the island of Hven, off the Danish coast, on which to build a new observatory. Tycho built a "Castle of the Heavens," Uraniborg, surpassing in glory any other observatory in the world at that time. It was equipped with an array of instruments of previously unobtainable precision, produced by the finest artisans Tycho could find. Tycho lived the life of a prince, entertaining visiting dignitaries in the grandest of manners. Income from tenants on the island kept the coffers full, to fund astronomy and his extravagant lifestyle.

Tycho dedicated fifteen years to measuring the positions of the stars and preparing observations for a catalogue of the heavens. When eventually printed this catalogue of stars was the first since that of Hipparchus. The motions of the planets were measured with increased precision. He recorded the passage of a great comet in 1577, demonstrating that its passage lay far beyond the Moon, making it an astronomical object rather than a meteorological phenomenon, which comets had previously been thought to be.

Tycho agreed with Rome that the Copernican universe was heretical. His principal scientific argument against the Copernican model was that he could not detect for stars the phenomenon of parallax. Tycho argued that if the Earth was orbiting the Sun, then parallax should be observed for the planets (even allowing for their motion) with respect to the distant stars when the Earth over a six-month interval moved from one side of the Sun to the other. Yet with the immaculate measuring equipment he had at his disposal, the most precise instruments the world of astronomy had ever known, no parallax could be observed. Thus, Tycho reasoned, the Earth could not be moving around the Sun—or if it was the stars must be at the very least seven hundred times farther away than the most distant planet. This latter notion was quite incomprehensible at the time. Hence, he argued, the Earth could not be in orbit around the Sun—it must lie stationary, at the center of the universe, with other bodies orbiting around it. Although Tycho challenged a simple interpretation of the Copernican theory, he also recognized that the Ptolemaic theories were not satisfactory either. Hence he invented his own strange hybrid system, whereby the planets orbited the Sun, and the Sun and Moon orbited the Earth, which was at the center of the universe (thus keeping the church happy). But no one other than Tycho could see any merit in this peculiar geometry.

Despite having an exceptional intellect and an inquiring mind, Tycho also believed in astrology. But this was a period when astrology still had a strange hold on scientists. (A century later, the great Isaac Newton would also be entranced by astrology, and alchemy.)

Given his unusual upbringing, Tycho had become used to extreme wealth and power, and his administration of Uraniborg was tyrannical. He was not liked by the observatory staff or by the Hven tenants, who were not impressed with their harsh landlord. One of the buildings at Uraniborg was a prison, where Tycho locked up tenants who failed to pay their rent and also staff and tenants with whom he had disagreements. Not surprisingly he made many enemies. But royal patronage meant his position was secure while the king of Denmark survived. However, with the death of King Frederick II there were many scores to be settled, and Tycho was eventually forced to leave his native Denmark. His magnificent "Castle of the Heavens" fell into disrepair and was soon a ruin.

Tycho moved to Prague, accepting the invitation of Emperor Rudolph II to become his court mathematician. But the emperor's main interest was astrology, and he was short of money so was unable to provide Tycho with the funds he needed to take forward his astronomical observations. Just two years after his arrival, Tycho died, never having been able to reestablish the grandeur and authority of Uraniborg. Nevertheless the quality of his observations of the stars, and the precision of his measurements of planetary motion, ensure him a place of prominence in the history books of science.

Shortly after arriving in Prague, Tycho had been joined by a young acolyte called Johannes Kepler. Kepler inherited Tycho's precision measurements of the planets and would put them to excellent use. Kepler had been born in Weil der Stadt in Germany in 1571, the year before the outburst of Tycho's new star. (His later astrological beliefs led him to claim, rather immodestly, that the 1572 new star was a portent of his own path to eminence in astronomy.) Kepler was a sickly child, inflicted with a crippled hand and poor eyesight. He studied at the university of Tubingen, expecting to enter the church. While there he came under the influence of Father Michael Mastlin, the professor of mathematics, who introduced him to the works of Ptolemy and Copernicus. Being a man of the church, Mastlin promoted the reality of the Ptolemaic doctrine and argued against the Copernican model. Kepler disagreed with him, recognizing the merits of Copernicus's ideas. Kepler accepted that his poor eyesight precluded a career in astronomy, and he moved initially to teaching. While teaching mathematics at Graz, he mastered the geometrical teachings of Euclid and turned them to a description of the motion of the planets. He published a book called *Mysterium Cosmographicum* that, while containing much doubtful astrology and no more than basic mathematics, so impressed Tycho when he read it that he invited Kepler to join him in Prague. He no doubt hoped that Kepler would be able to legitimize his own hybrid model for the cosmos, but Kepler was never impressed with it.

The meeting of Tycho and Kepler was of great fortune for astronomy, since Tycho bequeathed to Kepler his valuable cache of planetary observations. In the hands of such a gifted mathematician, they would be turned into three fundamental laws of planetary motion that would help bring order to the uncertain nature of the heavens.

With Tycho's death so soon after Kepler joined him, Kepler

gained the post as imperial mathematician to emperor Rudolph II. Being essentially a theoretician, he was less demanding of funding from the emperor than Tycho had been, and was willing to feed his patron astrological interpretations. But he was able to complete some astronomical observations of note, being fortunate to witness another brilliant outburst of a new star (another supernova within the Milky Way, it is now realized), which blazed forth in 1604. Again Kepler was keen to attach astrological connotations to the new star's appearance.

Kepler's careful analysis of the Hven observations enabled him to derive his three laws of planetary motion. The first law says that planets orbit the Sun in elliptical orbits, with the Sun at one of the foci. Kepler recognized that the planetary orbits departed from circular orbits only slightly, but even just a slight eccentricity (rather than the perfectly circular orbits Ptolemy, Copernicus, and Tycho had insisted on) allowed Tycho's observations to slot perfectly into place. The second law states that the radius joining a planet to the Sun sweeps out equal areas in equal times. And the third law states that the square of the planet's period of orbit around the Sun is proportional to the cube of its mean distance from the Sun. We do not need to contemplate the nature of these laws at this stage. Suffice it to say that they secured the reality of the Copernican revolution, which had moved the Earth from the center of the universe; and sixty years later, in the hands of Isaac Newton, they would lead to the universal law of gravitation—the basic tool in unraveling the true nature of the cosmos.

Despite his advocacy of the ideas of Copernicus, Kepler seemed to escape the hostile attention of the church. Perhaps its leaders failed to understand the full implications of his complex mathematical presentations.

Kepler's third law was described in his book *The Harmonies of the World* (a Pythagorean concept, it will be recalled), which, like so many of his writings, was a strange amalgam of carefully argued scientific and mathematical logic, and astrological bunkum. The astrology no doubt kept his patron happy. But life for the emperor, and Kepler, was becoming difficult. The emperor was forced to abdicate. And the state refused to make good Kepler's salary arrears for his position as imperial mathematician. Without a wealthy patron Kepler had to move to Linz in Austria to find a teaching post. His wife and one of his children died. On top of these personal tragedies his mother was charged with witchcraft, and Kepler's fight to have her acquitted was a serious distraction from his scholarship. He was forced by local religious strife to leave Linz in 1626. But he managed to press on with perfecting the analysis of Tycho's data and eventually published a set of tables predicting planetary motion based on them. He called the publication the *Rudolphine Tables* (in belated honor of his initial patron, Emperor Rudolph). Almost destitute financially, he traveled back to Prague, still hoping to recover the arrears of his salary of many years earlier. The journey proved too much for him, and he died on November 15, 1630. His planetary laws, the *Rudolphine Tables* of planetary motion, and his supernova of 1604 ensured that his name would be enshrined in textbooks of astronomy for generations to come.

Galileo Galilei (usually just called Galileo) is famed as the "father" of the astronomical telescope, and much else besides. He was born in 1562 in Pisa, Italy, and received his early tuition privately. He then moved to a monastery until 1581, when he returned to Pisa to study medicine. He was more attracted to mathematics and physics than medicine and left the university without a medical degree. His gift for novel experiments was revealed early, and he showed that the period of a pendulum depended on its length rather than the amplitude of oscillation—apparently by using his pulse to time the swing of a lamp in Pisa Cathedral. In 1589 he became professor of mathematics at Pisa University. In his most famous experiment he dropped two cannonballs of unequal weight from the Leaning Tower of Pisa, demonstrating that unequal weights fall at the same speed—an experiment that would be demonstrated to fascinated television viewers by U.S. astronauts on the Moon almost four hundred years later (albeit with a feather and a hammer, since on the Moon's surface air resistance was not a factor that needed consideration).

In 1592 Galileo moved to Padua to become professor of mathematics. Over the next seventeen years he formalized his ideas on the nature of falling bodies and acceleration.

In 1609 Galileo produced his own telescopes, based on reports of the recent invention in Holland by a Dutch lens maker, and turned them on the heavens. This type of telescope using lenses is called a refracting telescope. The main lens of a refracting telescope, called the object glass, can collect dramatically more light than the unaided human eye, bending the light so that it is brought to a focus to be viewed by an eyepiece lens and, depending on the design of the eyepiece, magnifying the image to a certain extent.

In 1610 Galileo published his book reporting the astronomical discoveries he had made with his new telescopes. The exciting new vista on the heavens caused amazement to his readers. He had discovered four moons rotating around the planet Jupiter and had also discovered the rings of the planet Saturn. By projecting an image of the Sun onto a screen, he discovered spots on the Sun. (Although sunspots had been recorded in abundance by the ancient Chinese, such apparent solar imperfections would have been discounted if sighted in Europe, because of the hold of the Aristotelian doctrine of perfect heavenly bodies.) He found that the spots drifted across the Sun, disappearing behind one limb and then reappearing on the opposite limb. Thus he argued that the Sun rotates about an axis. He saw mountains and craters on the Moon—a wondrous sight to behold. And he found that the nebulous band of the Milky Way was made up of closely packed stars, which were unresolved with the unaided human eye.

Galileo pressed the case for using the method of parallax to estimate the distances to the stars. He realized that since his telescope failed to magnify stars—they still remained as unresolved points of light regardless of the magnification of his telescope—they must be enormously distant. He suggested that two stars lying close together on the sky should be used for parallax measurement, so that many of the distorting effects faced by astronomers due to the presence of the atmosphere would cancel out.

The moons of Jupiter looked like a miniature solar system; and when Galileo discovered the phases of the planet Venus, which could be explained only in terms of its orbiting the Sun, he became convinced of the reality of the heliocentric model of Copernicus. He expounded its reality with vigor and, not surprisingly, soon found himself in trouble with the church as a consequence. He sought to smooth out his differences with the church, however, and thought he had succeeded.

In 1632 Galileo published a book called *Dialogue Concerning the Two Chief World Systems*, in which he presented three fictitious individuals debating the relative merits of the Ptolemaic and Copernican universes. The pope felt that the bumbling fool Simplicius, the character Galileo used to present the Ptolemaic view, might have been based on himself. The Sacred Congregation of the Index placed the book on the forbidden list—and there it remained until a slightly more enlightened Catholic Church agreed to its removal in 1835.

Galileo was taken to Rome to face trial in front of the Inquisition, and under the threat of torture he was forced to recant. His humiliating climb down meant that he escaped imprisonment and was allowed to return to his villa near Florence. But he was forbidden from doing any further work in astronomy and was placed under virtual house arrest. Nevertheless no one could control his love of science, and he did undertake many important experiments in his final years. He published the results of his experiments in a new book titled *Discourse and Mathematical Demonstrations Concerning Two New Sciences*, in which he refuted Aristotelian mechanics. In his final years his sight failed, and he died in 1642 a broken and bitter man. But few individuals have done as much for science as did Galileo. He demonstrated the power of experimental verification of theoretical hypotheses. And his telescopes brought a new capability to the study of astronomy. The heavens could withhold their secrets no longer.

Faintness clearly implied that the stars, if they were in any way like the Sun, had to be at extreme distances. The great Dutch physicist Christiaan Huygens (who had been born in 1629) decided to try to compare the brightness of Sirius, the brightest star in the night sky, with that of the Sun. He developed a clever technique to attempt this. It involved letting sunlight pass into a darkened room through a small hole and adjusting the hole size until the light transmitted through it matched, as far as Huygens could estimate, the brightness of Sirius. Then he calculated the angular size of the hole as a fraction of the angular size of the Sun. Then, using the inverse square law for light (such that if the distance of a light source is doubled then its observed brightness will drop to a quarter), he suggested that Sirius must be at a distance of nearly twenty-eight thousand times the distance to the Sun. Although he had underestimated by a factor of twenty, Huygens found the distance he had derived difficult to comprehend:

> For if 25 years are required for a bullet out of a cannon, with its utmost swiftness, to travel from the Sun to us then by multiplying the number 27,664 into 25 we shall find that such a bullet would spend almost seven hundred thousand years in its journey between us and the nearest of the fixed stars. And yet when in a clear night we look upon them we cannot think them above some miles over our heads. What I have here enquired into is concerning the nearest of them. And what a prodigious number must there be besides of those which are placed so deep in the vast space of heaven, as to be as remote from these as from the Sun! For if with our bare eye we can observe above a thousand, and with a telescope can discover ten or twenty times as many, what bounds of number must we set to those which are out of reach even of these assistances! Especially if we consider the infinite power of God

Astronomers were starting to get the feel of a cosmos of staggering proportions. But now the true master of the universe was about to make his move.

Arguably the greatest scientist until the modern era was Isaac Newton. He was born in 1643 at Woolsthorpe, England. After a schooling oft interrupted by illness, Newton's ability was recognized by an uncle who encouraged him to go to university. He entered Trinity College, Cambridge, in 1661. In 1665 the university was closed because of the plague, and he returned home to Woolsthorpe. Provided with a period for quiet contemplation, he formulated at this time many of his ideas on gravitation and the nature of light, and he developed the mathematical technique that would eventually become the calculus. Supposedly while sitting under an apple tree in the garden at Woolsthorpe, he surmised that the gravity that caused an apple to fall could extend as far as the Moon and keep it in its orbit around Earth; and if gravity could extend to the Moon perhaps it could extend indefinitely—sowing the first intellectual seeds for a universal law of gravitation. His optics experiments, carried out in the darkened study at Woolsthorpe by passing light through a prism, demonstrated that white light is made up of the superposition of light of all colors—a result that would later give birth to the science of spectroscopy, which would prove of great value in studies of the scale of the cosmos.

Newton returned to Cambridge in 1666, when the university reopened after the plague, and he became the Lucasian Professor of Mathematics at the age of just twenty-six years. He was elected a Fellow of the Royal Society in 1672. He remained at Cambridge for thirty years. His lectures failed to inspire his students—yet his natural philosophy was profound. During this time he perfected the work that would set him aside as one of the greatest of scientific minds, producing his laws of motion, revealing the nature of gravity, making fundamental discoveries in optics (including inventing the reflecting telescope), and formulating the calculus. Edmund Halley, a wealthy man at the time (although a family fortune would later be lost) and a friend of Newton, sponsored the publication of Newton's greatest work—the *Philosophae Naturalis Principia Mathematica* (usually known merely as the *Principia*).

Many scientists had tried to perfect a reflecting telescope before Newton brought his skills to the task. Newton was the first person to build a reflecting telescope that worked, and it was this achievement that earned him his Royal Society fellowship. Rather than using a lens to collect the light from a faint object, as in the refracting telescope, a reflecting telescope uses a concave mirror to bring light to a focus. The image can then be seen through an eyepiece at the focus. But Newton, in his telescope, used an angled plane mirror near the focus to divert the light to an eyepiece mounted on the side of the telescope. In more modern reflecting telescopes, in a so-called Cassegrain arrangement, a convex mirror near the focus of the concave mirror redirects light back through a central hole in the concave mirror for viewing. The larger the mirror of a reflecting telescope, the more light it can collect (in the same way that a rain barrel with a large opening at the top will collect more rain than one with a small opening). Fainter objects can thus be detected the larger the telescope's mirror. But the larger the mirror, the greater the challenge to produce it with the accuracy required to bring the light to a perfect focus.

Newton's three laws of motion clarified the nature of forces (a force is a "push" or a "pull"). The first law states that any object remains at rest, or in its state of uniform motion, unless acted on by an external force. The second law states that if a force acts on an object, the object will be accelerated along the direction of the force. (The second law led to the concept of "momentum," which is the product of the mass of any object and its velocity.) The third law states that for every force that acts on an object, there is a second force equal in magnitude but opposite in direction that acts on a second object. The three laws would help describe the behavior of celestial objects.

The *Principia* included the formulation of the universal law of gravitation. Its publication arose from an intriguing interplay between Newton in Cambridge and a cohort of establishment scientists in London. Halley had had regular discussions with two well-known London scientists, Robert Hooke and Christopher Wren, on the nature of gravity holding the planets in orbit around the Sun. (Wren and Hooke were collaborating in the rebuilding of London following the great fire of 1666, but as Fellows of the nascent Royal Society, and polymaths by nature, they found time to speculate on the wonders of nature.) When Halley suggested that the elliptical orbit of a planet around the Sun might be explained by gravitational attraction varying as the inverse of the square of the distance from the Sun to the planet (a so-called inverse square law), Hooke boasted that he had already developed a mathematical proof demonstrating just that fact. Despite the encouragement of his colleagues, however, he could never produce a copy of the mathematical proof he claimed that he had achieved. A skeptical Halley decided to visit Newton in Cambridge to ask what he thought, since it was known that Newton was developing new ideas on gravity. Newton (who would become an archrival of Hooke) said that he himself had a mathematical proof that a force obeying an inverse square law would result in elliptical orbits, as with the planets. He later sent Halley a copy of the written proof, which would eventually become part of the *Principia*. His work was based on ingenious geometrical proofs deduced from Kepler's laws of planetary motion, and although Newton's inquiry was limited to the solar system, he concluded that the gravitational force was universal. His law states:

> Every particle in the universe attracts every other particle with a force that is directly proportional to the product of the masses of the two particles and inversely proportional to the square of the distance between them.

In 1689 Newton was elected a member of Parliament; in 1699 he became master of the Royal Mint. But now his prodigious scientific output was beginning to diminish, although he was still very influential in scientific circles and produced a monumental treatise on light (based on his earlier research) and two new editions of the *Principia*. He was elected president of the Royal Society in 1703, holding the post until his death in 1727. During this period he held sway over the development of science in England, and his acolytes held all the positions of note in British science while his enemies were ostracized from the mainstream of science. Conflicts with other scientists, an intense sensitivity to criticism, and a misguided possessiveness over his ideas marred an otherwise illustrious scientific career. His dispute with the German philosopher Gottfried Leibnitz over the origin of the calculus was one of the greatest disputes of priority in the history of science. Newton hounded Leibnitz to his grave, and even in death denounced his memory. Despite these traits of character revealing a flawed genius, and a peculiar fascination with alchemy and astrology, Newton's contributions to the establishment of science have no parallel.

In the century from Copernicus to Newton, the understanding of the universe had been transformed. The Earth had been firmly dislodged from its position of celestial preeminence at the center of the Ptolemaic universe. The nature of the orbits of the planets had been revealed by the masterful observations of Tycho and their ingenious interpretation by Kepler. The failure to detect parallax in stars was accepted as an indication that they must be at vast distances beyond the solar system, such that any parallax would be too small to be measurable with current instruments. Galileo introduced the telescope and produced observations that provided validation of the new ideas. And the genius of Newton brought forth the reflecting telescope, new laws of motion, and an understanding of the fundamentals of optics. It also delivered the theory of universal gravitation, which explained the motions of the planets and identified the primary force in shaping the universe. Real science now had its firm foundation.

An understanding of the heavens was of more than scholastic interest. The stars were useful for navigation, and powerful seafaring nations realized that improved navigation meant control of the oceans. While latitude (the position of a ship north or south of the equator) could be estimated with precision from measurements of the elevation of the stars, estimating longitude (the position east or west from some reference point) was more problematic since it required knowledge of the time at the reference point. In the absence of artificial timepieces that could withstand both the harsh conditions and the length of journeys at sea, improved astronomical observations offered the most reliable way to measure time and, therefore, longitude. A solution was needed urgently: simply put, the nation that solved the riddle of longitude could control the oceans.

This sense of purpose propelled both the French and the English monarchies to found royal observatories in the late seventeenth century. The first three successive English royal astronomers—John Flamsteed, Edmund Halley, and the Revered James Bradley—particularly distinguished themselves as cataloguers of the heavens. By the time of Bradley's death in 1762, the most up-to-date catalogue of the sky would contain observations on the positions of some sixty thousand stars, the movement of the planets, the passage of comets, and the meandering path of the Moon.

But though astronomers now had a reliable catalogue of the heavens, laws to describe the motion of the planets, a universal law of gravitation, and telescopes of increasing sophistication, they still did not know how far away the stars lay. And without a cosmic yardstick, the heavens would always be just two-dimensional: a flat vision of the sky, devoid of depth, shape, and character. Distances were needed to make sense of the cosmos. As royal astronomer, Bradley made the problem of the stars' distance the central problem of the Royal Observatory.

Bradley hoped to use parallax to complete his calculations. In 1671, French astronomers had used this method to estimate the distance to the planet Mars. One team was based at the observatory at Paris, while their colleagues journeyed to Cayenne in French Guiana. They made simultaneous measurements of the position of Mars with respect to the background stars. Using the then best estimate of the distance between their two observing points, they were able to calculate the distance to Mars. Armed with Kepler's laws, and knowing the time it took each planet to orbit the Sun, they could then estimate the distance each planet lay from the Sun. Their estimate for the distance from the Sun to the Earth was 140 million kilometers—not so different from the modern-day estimate for the average distance between the Earth and the Sun of 149,597,870 kilometers. (The distances to the nearest planets have been found with remarkable precision in modern times by bouncing radar signals off them and measuring the time required for the radar pulse to get there and back.)

Bradley knew that measuring stellar parallax would be exceedingly difficult. All previous attempts had failed, implying that the stars were at incredible distances. Even Robert Hooke, noted for experiments of great ingenuity, had failed to find parallax when he set up a special telescope pointing vertically though two stories of his house to look at the star Gamma Draconis, since it conveniently passes directly overhead at the latitude of London. Bradley had a special instrument developed for his parallax experiment and erected at Kew. He also chose the star Gamma Draconis to measure. To his initial delight, Bradley was able to measure small angular shifts, but with further observation his elation turned to puzzlement. Over a year the star appeared to trace out a small circle in the heavens. Measurements of other stars demonstrated similar behavior. This was not what was expected from the simple parallax phenomenon.

Bradley thought of an explanation for the strange behavior in a most unexpected way. While sailing on the Thames, Bradley noted that the weather vane on the masthead indicated a slightly different direction for the wind when the boat changed the direction it was heading. Yet it was clearly not plausible to think that each time the boat changed its heading the wind made a subtle change in direction simultaneously. The effect is in fact due to the relative motion of the boat and the wind. (Consider the analogy of raindrops falling vertically down the window of a stationary car; when the car is moving, the raindrops no longer fall vertically relative to the car but at an angle that depends on the speed of the raindrops and the speed of the car.) Bradley envisaged Earth as being like the boat and starlight as being like the wind. The subtle change in apparent direction is known as "aberration." When Earth is moving directly toward or away from a star during its passage around the Sun, then no aberration is detected; but the aberration effect is most pronounced every six months when Earth is moving at right angles to the direction of the star.

With aberration Bradley had made a discovery of singular importance, but he had not discovered parallax. He need not have feared for his reputation—the search for parallax would defeat brilliant minds for yet another century.

There was an important discovery at the end of the seventeenth century that is worth reporting here. The moons of Jupiter were proposed as one possible astronomical clock for use in longitude measurements. The time at which they would pass behind Jupiter could be predicted, and when such eclipses were observed, then the time back at port (and hence longitude) could be estimated. In truth eclipses of the moons of Jupiter were too difficult to observe from the tossing deck of a ship, so the method was never of any practical use. However, in trying to perfect the technique, a Dane called Ole Römer, working at the Paris observatory, tried to understand the peculiar feature that eclipses of Jupiter's moons sometimes occurred earlier than predicted—and at other times later than predicted. It was discovered that eclipses occurred too early when Earth was closest to Jupiter, and they occurred too late when Earth was farthest from Jupiter. Römer figured that the reason must be that light travels at a finite speed. When Earth is farthest from Jupiter the light from the eclipsing moon takes longer to reach us, and therefore the eclipse appears to be taking place late. From such observations Römer estimated that light must travel at a finite, albeit enormous, velocity of some 186,000 miles each second. This was a startling insight, since previously it had been widely believed that observers saw light instantaneously however far away they were. Römer realized that the fact that light traveled at a finite speed meant that in looking out into the heavens one was also looking back in time—viewing a light source at the instant its light had started its journey, sometime prior to its being sighted. The difference in eclipse timings gave him an estimate for the diameter of Earth's orbit around the Sun in light-minutes.

William Herschel is acknowledged to have been one of England's greatest observational astronomers. He was a German by birth but emigrated to England in 1757 from Hanover, where he had been an oboist in the Hanover Guards. He remained a professional musician in his newly adopted land but pursued astronomy as a hobby. While doing so he discovered a new planet in 1781, which later was named Uranus. This single discovery gained him the patronage of King George III (the king who "lost" the American colonies but who was a keen patron of science), enabling Herschel to take up astronomy as a full-time occupation.

Like Bradley, Herschel made an unexpected discovery in searching for parallax. He selected for observation stars close to each other in the sky, following the guidance of Galileo. Although Herschel failed to find parallax, he did find that in some cases the stars he observed appeared to be in relative motion around their common center of gravity. They were clearly "binary stars," bound by gravity to orbit one another. It is now realized that almost half of all stars can be found in binary systems.

After Herschel's death three astronomers, working independently, took up the search for the elusive parallax. The first was

Friedrich Bessel, director of the Konigsberg Observatory in Germany. The second was Friedrich Struve, director of the observatory at Dorpat in Estonia (by then ruled by Russia). The final parallax hunter was Thomas Henderson, working at the British southern observatory set up at the Cape of Good Hope. The race was on.

Bessel started his search in 1837 with a binary star system in the constellation of Cygnus, numbered 61 in Flamsteed's catalogue. Within just a year he announced that both components of 61 Cygni showed a parallax of a mere 0.3 seconds of arc, when observations were taken six months apart with Earth on opposite sides of its orbit around the Sun. There are 60 seconds of arc in a minute of arc, and 60 minutes of arc in one degree—so 0.3 seconds of arc is tiny. Indeed 0.3 seconds of arc is about 0.0001 degree (one ten-thousandth)—a truly minute angle, demonstrating why it had been impossible for Tycho, Hooke, Bradley, Herschel, and others to measure parallax with the instruments they had available to them, however much care they had taken. The diameter of Earth's orbit around the Sun was then taken as the baseline to estimate the distance to 61 Cygni. Bessel calculated its distance to be a staggering 11 light-years.

Parallax gives an alternative measure of astronomical distance to the light-year. For an imaginary star with parallax of one second of arc, its distance is defined as the parsec. One parsec equals 3.2615 light-years.

Although Bessel got his announcement in first, in fact the first actual parallax measurement had been achieved by Henderson. Thomas Henderson had been born in Dundee in Scotland in 1798. He was appointed the royal astronomer at the Cape in 1832, and during a brief thirteen-month stay there he took measurements of the bright southern star, Alpha Centauri. Because of ill health he was forced to return home to Britain, and he did not finish the analysis of his results immediately. It was not until after Bessel's announcement of the 61 Cygni distance that Henderson completed his calculations and demonstrated that Alpha Centauri was at a distance of 4½ light-years. But in science priority is determined by the date you submit your results for publication—not when you get around to analyzing your data. So Bessel is rightly recognized as the first person to measure parallax.

Friedrich Struve had the most difficult measuring task of all, since the star he had chosen was Vega. Its angle of parallax was smaller even than Bessel's, and the distance to Vega was found to be some 26 light-years. Although Struve was the last of the pioneers to detect parallax, he had the satisfaction of measuring the smallest parallax angle up to that time and hence the most distant star.

Bessel, Henderson, and Struve had at last broken the parallax jinx, demonstrating the incredible distances to even the nearest stars and finally bringing an experimental verification to the supposed immense size of the cosmos.

Parallax observations can be used only for relatively nearby stars. Once one gets beyond a distance of about 150 light-years, the minute angles are difficult to measure with certainty. There was another method developed to estimate star distances, which was called statistical parallax. It depended on the realization that stars are in a systematic motion around the center of the Milky Way. It is possible to measure the component of a star's motion along the line of sight to the star from the Doppler effect. And it is possible to measure its motion across the line of sight, its so-called proper motion, by taking images of the sky spaced over many years. Then it is possible to combine the radial velocities and proper motions statistically to give the expected orbital behavior around the center of the Milk Way. Of course individual stars may be in a random motion not perfectly in tune with the general orbital drift, because of the local gravitational effects of nearby stars. Thus the technique is of no use to find the distance to an individual star, since we are looking only at an average effect. However, the method can be of use in estimating the distance to a group of stars.

Distances in the cosmos are determined via a number of overlapping steps, with each step depending on the precision achieved in the preceding one. An error in any one step will negate the accuracy of successive stages, hence the need to develop complementary methods to gain confidence in the distance estimates at each stage. But the critical first step had been made with trigonometric parallax measurements for the nearby stars. Now the way would hopefully be open to measure distances out to the extremities of the cosmos.

It had taken two and a half millennia—from the imaginative philosophies of the ancient Greeks, through the scholarship of the Arabs, to the collective genius of the post-Renaissance European astronomers—for a true understanding of the distances to the stars to be achieved. But with the arrival of the twentieth century, and the advent of new technology, spectacular advances in understanding the true scale of the cosmos would be achieved in a mere two and a half decades. This new understanding would be based on controversy and intellectual conflict that would divide the world of astronomy.

Chapter 3

THE GREAT DEBATE

It is rare in the history of human understanding for a new epoch to be defined by a debate. One such occasion was on April 26, 1920, when the U.S. National Academy of Sciences arranged a meeting in Washington, D.C., in an attempt to resolve the questions of the scale of the universe and the nature of the spiral nebulae. There was controversy aplenty surrounding the issue, since uncertainty remained about the extreme distances to the stars and whether the universe extended beyond the Milky Way or whether the Milky Way was itself the whole cosmos. Many of the most eminent scientists of the day held the latter, more traditional, position, but a few imaginative astronomers had begun promoting the so-called island universe theory, which reduced the Milky Way to a mere part of a much larger system—albeit without much hard evidence. The situation had reached a crisis point because pressing astronomical questions about the size, mass, and age of the universe could not be answered until this issue was resolved.

To understand why spiral nebulae occupied such a central role in early-twentieth-century astronomical debates, we must return briefly to William Herschel's research over a century earlier. One of Herschel's principal interests was to determine the structure of the Milky Way. He was assisted in this task by his sister Caroline, a talented astronomer in her own right who won acclaim as a discoverer of several comets.

The method William Herschel used to investigate the structure of the Milky Way he called "star gauging." He made the simplify-

ing assumption that all stars shine with the same inherent brightness, as if they are all identical candles. On this assumption the
fainter a star, then the farther away it must be, following the well-
known inverse square relationship. Herschel's star-gauging method
involved counting the number of stars in several directions, assuming a uniform star density, and hence estimating the distances
needed to count so many stars in these directions. Thus along his
chosen directions he could estimate the relative extent of the universe (so long as it was uniformly populated with stars). This was a
truly monumental task, taking Herschel and his sister twenty years.
By the time they had finished they had counted over ninety thousand stars in 2,400 sample areas. This was one of the heroic surveys
in the history of astronomy. On the basis of star gauging Herschel
concluded that the universe was a flattened disk, like a millstone,
with a diameter at least six times its thickness. (What Herschel
could not know was that stars are not standard candles and that dust
that lies along the plane of the Milky Way dims the brightness of
more distant stars, both of which affected his results. Nevertheless
his flattened disk interpretation was indeed correct.) While William Herschel's assumption that all stars shone with the same intrinsic brightness did not stand the test of time, his belief that standard candles could be used to measure distances and determine the
structure of the Milky Way would eventually find a successful application a century and a half later.

During their studies of the Milky Way, the Herschels pondered
the nature of small, faint, light-emitting clouds called nebulae (from
the Latin for clouds). The famous French comet hunter Charles
Messier had prepared a catalogue of nebulae, which might cause
confusion in his searches for comets. The Herschels' study of the
Milky Way had revealed some 1,500 more such objects. Herschel
initially pondered that they might be separate galaxies like the
Milky Way but lying at great distances from it, so that it was difficult to resolve their individual stars. This brilliantly intuitive guess
made little impression at the time, despite its coincidence with the
island universe theory of Immanuel Kant. Herschel believed that
his telescope could just resolve stars in twenty-nine of the nebulae
he had studied, but he then found many other nebulae as large or

larger in which no stars were evident even though, on the basis of their size, their stars might have been expected to be resolved. In fact Herschel himself helped to undermine the island universe idea through his work on certain gaseous disks called planetary nebulae, which showed no evidence of containing stars. In 1811 he wrote: "We may have surmised nebulae to be no other than clusters of stars disguised by their very great distance, but a longer experience and better acquaintance with the nature of nebulae, will not allow a general admission of such a principle."

The idea that the nebulae could be planetary systems in the making within the Milky Way gained popularity in the nineteenth century. But still many astronomers and philosophers pondered whether at least some of the nebulae might actually be distant independent galaxies—the "island universes" of Kant. In the mid-nineteenth century the third earl of Rosse (an Irish nobleman who was one of the great early builders of giant telescopes) built a very special telescope at his home at Birr Castle. It used a 72-inch mirror to gather light from very faint objects. He discovered that many of Herschel's resolvable nebulae were distinguished by a characteristic spiral form. The spiral nebulae became a source of fascination to astronomers, with opinion soon seriously divided on their nature. It would take a further seventy years for the nature of the spiral nebulae to be resolved, and they remained a source of intense interest and diverse explanation.

Such was the intellectual context in 1920 when many of the nation's scientific elite gathered to witness a debate on the nature of the Milky Way. Although the organizers of the event in Washington, D.C., did understand that a meeting to look at the questions was particularly timely, it is unlikely that they would have anticipated that it would define a major new era in the understanding of the universe. In the annals of astronomy the occasion has become known as the Great Debate. In truth few of those who attended would have found it "great," even if they had been intrigued and entertained. Nor was it a "debate" in the strictest definition of the term, involving the cut and thrust of opposing ideas and the challenging of evolving arguments. There were no interjections—no points of order—none of the passion and emotion of memorable political debates. But no matter. Inspired by the issue at the heart of the meeting, astronomers resolved to secure the evidence needed to determine the nature of the universe with certainty. The debate would generate an intense rivalry between two scientific camps: one headed by an ambitious young astronomer named Harlow Shapley from the Mount Wilson Observatory, and the other by an "old guard" astronomer named Heber D. Curtis from the Lick Observatory. Not only was this a clash between scientific ideas; it also pitted old against young—and radicals against conservatives. Two rival observatories were striving for supremacy.

The advances stimulated by the Great Debate would represent a heroic victory for the human spirit—ironically by making us aware of our insignificance within the universe. The whole episode would demonstrate how science can advance through an adversarial process, with different groups adopting extreme positions so that a true understanding eventually emerges as controversies are addressed and finally resolved. The debate was in the true tradition of famous scientific controversies as characterized by the Darwinism debates at Harvard University and the Relativity debates of the Royal Society in London.

The Washington of the years immediately following the First World War led a nation that was uncertain how to react to a victory in a war that many considered had not been their own. Woodrow Wilson had championed a new world order under the custodianship of the League of Nations but could not persuade a Congress attracted by isolationism to join the league. U.S. industry was responding with vigor, new technology, and efficient production lines to a postwar demand that the traditional industrial powerhouses of Germany, France, and Britain could not meet as they recovered from the exhaustion of a mindless conflict. U.S. national self-confidence was high and would lead to the socially liberated twenties, with wild financial excesses that ended in the Wall Street crash of October 1929. In 1920 the Nineteenth Amendment to the U.S. Constitution gave the women of America the vote for the first time. Herbert Hoover's "Noble Experiment" of Prohibition had been introduced in early 1920, but in fact it led many citizens to flout the law. The Roaring Twenties, a decade known through its images of flappers, jazz, and jalopies, were fueled by bootleg gin consumed in speakeasies to which corrupt cops turned a blind eye.

The world of science still involved a small intellectual elite in 1920, but it was coming of age. In no field of scientific endeavor was this more apparent than in astronomy, and in particular in understanding the nature, true extent, and origin of the universe. A powerful new generation of U.S. astronomical telescopes, funded through the patronage of enlightened industrial barons, could surely provide new insight into the true scale of the universe. The progress in astronomy was going to be a partnership between European theoretical creativity and New World wealth and observational ingenuity. It would bring forth a generation of scientists from a variety of backgrounds, ranging from university professors to a crime reporter and at least one mule-train driver.

Although the meeting of the National Academy of Sciences in Washington in April 1920 took on a status of importance in the annals of science, its origins were inauspicious. The idea for the debate came from George Ellery Hale, founder and first director of the Mount Wilson Observatory in California. Hale was the greatest of all astronomical telescope builders, without whom the race to measure the cosmos could not have been run with such vigor. Late in 1919 Hale, a prominent member of the U.S. National Academy of Sciences, had proposed that an upcoming meeting in Washington should be devoted to a debate. The event would be paid for from a fund set up in memory of his father, William Ellery Hale, who had made a fortune building elevators in Chicago in the construction boom that followed the great fire in 1891. Hale elevators became world famous and even graced the Eiffel Tower in Paris. Hale senior funded a superbly equipped private observatory for his only child, George Ellery, who developed a boyhood passion for astronomy. When his son became a professional astronomer Hale senior generously supported and encouraged him. George Ellery's private observatory was impressive, even by professional standards.

George Ellery Hale took his degree at the famous Massachusetts Institute of Technology (MIT). During his sophomore year he persuaded the director of the Harvard College Observatory to take him on as a volunteer assistant; he was to prove himself an adept persuader throughout his career—especially when it came to extracting money from people to build telescopes. In 1892 he joined the new University of Chicago and established an observatory there. This was the first of many successful fund-raising attempts, all in the name of building state-of-the-art telescopes. In this case he found a patron in a trolley car tycoon, Charles T. Yerkes. Yerkes had made a fortune constructing the Chicago elevated railway. Although Yerkes was reluctant to part with his cash, his gift was acknowledged by having the observatory bear his name thereafter.

In 1894 Hale launched the *Astrophysical Journal*, which would in time become the preeminent journal for professional astronomers. In 1904 he moved to Mount Wilson in California (in part the move was motivated by the warmer climate to assist the rehabilitation of a sickly daughter); he established a leading observatory there with its famous 60-inch telescope. (This measurement refers to the diameter of the mirror collecting light within the telescope; the larger the mirror, the fainter the astronomical objects the telescope can detect.) Hale raised the funding for new telescopes at both the Yerkes and Mount Wilson observatories, and was a leading force in the vision for a giant 200-inch telescope built later at a new observatory at Palomar Mountain in California. Although he had been born "with a silver spoon in his mouth," which would have enabled him to pursue a life of leisure, and drew heavily on his family's fortune and political connections throughout his career, Hale's energy and enthusiasm for building increasingly sophisticated telescopes was impressive by any measure. He was active in the politics of science, favoring international collaborations. Hale was instrumental in the interwar establishment of the International Astronomical Union (since all the previously existing international scientific organizations had been disbanded under the terms of the Treaty of Versailles). George Ellery Hale was an intellectual giant of the age.

Hale had suggested two possible topics for debate at the proposed meeting in Washington. The first topic was relativity. Relativity was the brainchild of the great Albert Einstein (and is a topic we will consider in greater detail later). In 1920 it was a subject of particular fascination and topicality, since earlier in 1919 predictions from Einstein's theory of general relativity that light would be bent in a gravitational field had been verified observationally. The second suggested topic was the "island universe" hypothesis. William Herschel had extensively studied the extended regions of emission, called nebulae. Some argued that the nebulae were merely clouds of gas lying within the Milky Way, while others tended toward the view that they were distinct massive star systems at distances so vast that their individual stars could not be discerned—the "island universe" hypothesis of Immanuel Kant.

Hale suggested that the case against the nebulae being separate island universes lying beyond the Milky Way should be presented by one of his colleagues from Mount Wilson, Harlow Shapley. William Wallace Campbell, director of the rival Lick Observatory, would, he proposed, present the case for the island universe hypothesis. The Lick Observatory was located in the Diablo Range east of San Jose in California. It had been funded in 1885 by an eccentric millionaire, James Lick, who had made an initial fortune from making pianos but then saw his wealth grow spectacularly through land acquisition in the Californian gold rush. The generous patronage of a variety of wealthy eccentrics played an important role in the emergence of the United States as the dominant world power in astronomy.

The secretary to the National Academy of Sciences, astrophysicist Charles Greeley Abbot, was not overly impressed with either of Hale's proposed topics for a debate and wrote to him on January 3, 1920:

> You mentioned the possibility of a sort of debate, either on the subject of the island universe or of relativity. From the way the English are rushing relativity in [the scientific journal] Nature and elsewhere it looks as if the subject would be done to death long before the meeting of the Academy, and perhaps your first proposal to try to get Campbell and Shapley to discuss the island universes would be more interesting. I have a sort of fear, however, that the people care so little about island

universes, notwithstanding their vast extent, that unless the speakers took pains to make the subject very engaging the thing would fall flat.

Hale was unmoved by this lack of interest in relativity, but Abbot's suspicions were unequivocal. He wrote to Hale again on January 20, 1920:

> As to relativity, I must confess that I would rather have a subject in which there would be a half dozen members of the Academy competent enough to understand at least a few words of what the speakers were saying if we had a symposium upon it. I pray to God that the progress of science will send relativity to some region of space beyond the fourth dimension, from whence it will never return to plague us.

In truth Hale himself was mesmerized by relativity, and confided:

> The complications of the theory of relativity are altogether too much for my comprehension. If I were a good mathematician I might have some hope of forming a feeble conception of the principle, but as it is I fear it will always remain beyond my grasp.

Although Abbot had reservations about both topics, but especially relativity, Hale continued to press his proposal for an island universe debate. He was noted for his persistence, and besides it was his family's endowment that was sponsoring the event. However, he now wanted Shapley's adversary to be Heber D. Curtis rather than Campbell. Curtis worked for Campbell and was an enthusiastic proponent of the island universe theory, so perhaps that is why Hale decided he would be a more worthy opponent. On February 18 Abbot confirmed his agreement to a debate in a cable to Hale:

> Am writing Heber Curtis suggesting Debate him and Shapley on subject scale of universe for Academy meeting forty

five minutes each suggest communicate Shapley and Curtis and wire if favorably arranged.

Harlow Shapley and Heber D. Curtis were both very eminent astronomers of their generation. Curtis, the older man, had a securely established reputation. Shapley was recognized as a young astronomer of enormous potential. Both were particularly interesting individuals, who had arrived at their positions of eminence in the scientific community from very different backgrounds.

The early career of Harlow Shapley, who was born in rural Missouri in 1885, hardly suggested that he would play a role in initiating a new cosmic revolution. He started as a crime reporter for a small-town newspaper in Kansas, covering the fights of drunken oilmen. Wishing to better himself, he entered the University of Missouri hoping to study journalism. He had not realized that the school of journalism had not yet opened, and once he discovered this he decided to investigate what other field of study might be selected. Shapley later described his choice in amusing prose:

> I opened the catalogue of courses. The very first course offered was a-r-c-h-a-e-o-l-o-g-y, and I couldn't pronounce it! I turned over a page and saw a-s-t-r-o-n-o-m-y. I could pronounce that—and here I am!

Shapley's path in astronomy was assured when he became a graduate student of Henry Norris Russell at Princeton University. Russell was an acknowledged prince among astronomers, one of the pioneers who brought together the classical methods of astronomy, physics, and spectroscopy into the new discipline of astrophysics. The later citation of a Yale honorary doctorate heralded Russell's stature in science: "Dean of American astronomers, profound student of the physics of the stars and the structure of the atom, master of the interpreting to us of the whole domain of astronomical research, philosopher of cosmic evolution and man's place in the universe."

Henry Norris Russell was quick to recognize the talents of the new graduate student who arrived to join him in 1912. He noted later that his research career attained a new impetus when "the Lord sent me Harlow Shapley." Russell became Shapley's adviser and friend.

After receiving his doctoral degree at Princeton in 1914, Shapley joined the staff of the Mount Wilson Observatory, working for George Ellery Hale. He was amazingly fortunate to gain a staff post at a leading observatory with almost unlimited access to telescope time throughout the war years. He turned his research interests to the study of groupings of stars called globular clusters. Globular clusters are immense, densely packed groups of stars within the Milky Way, some containing as many as a million stars. Using a newly developed technique for estimating the distance to variable stars (which we will discuss in some detail later), Shapley showed that globular clusters appeared to be distributed in a giant sphere centered in the constellation Sagittarius. Shapley argued that the center of the sphere, which he estimated was a staggering 100,000 light-years in diameter, was the center of the Milky Way. Such distances were almost impossible to comprehend at the time. Shapley's new estimate for the diameter of the Milky Way was some ten times that of conventional wisdom. So large was this new estimate of the Milky Way that Shapley and others argued that it represented the extent of the entire cosmos. They reasoned that there was nothing significant beyond the outer bounds of the Milky Way, which they claimed was large enough to embrace all the important astronomical objects. Many astronomers agreed that nothing else could possibly be needed beyond the bounds of such an enormous Milky Way to explain the totality of the universe.

Prior to Shapley it had been assumed that the solar system lay near the center of the Milky Way. This conventional picture of the solar system at the center of the Milky Way resulted from the observations of an eminent Dutch astronomer, Jacobus Cornelius Kapteyn. Working in the late nineteenth century, Kapteyn had refined Herschel's "star gauging" technique. Kapteyn's work had provided evidence that the Milky Way was a disk some 30,000 light-years across and 6,000 light-years thick, with the solar system lying close to its center (thus confirming Herschel's general picture of the Milky Way). The scale of the Kapteynian universe was impressive enough—a universe of vastness previously unimagined and still difficult to conceive of, even acknowledging Kapteyn's reputation. Yet now Shapley wished to expand the proposed dimensions of the Milky Way many times—and relegate Earth to its outer neighborhoods, some 30,000 light-years from its center. Here was what appeared to be the ultimate post-Copernican humiliation. God's Earth had been relegated from the center of the pre-Copernican cosmos to the outer reaches of the post-Shapley Milky Way. Shapley, the young, ambitious, rising star of astronomy, was making sure he was significant by asserting Earth's insignificance in the grand order of the cosmos.

Heber Doust Curtis was born in Muskegon, Michigan, in 1872, the son of a one-armed Union veteran called Blair Curtis and his wife, Sarah Eliza Doust. Heber Curtis studied classics at the University of Michigan and then followed a career as a schoolmaster—teaching Latin in the Detroit High School and then teaching Greek and Latin at Napa College, California. Napa had a small telescope, and Curtis became fascinated by astronomy. When Napa merged with the College of the Pacific, he transferred to become professor of mathematics and astronomy at the joint institution. In 1902 he managed to gain a staff appointment at the Lick Observatory. Apart from two years teaching navigation and conducting research in optics in support of the war effort, he remained at Lick until 1920. In 1910 he had been placed in charge of Lick's program for studying nebulae. He later confessed that he considered this work his most important contribution to astronomy. He was noted for his methodical and cautious approach, and he came from the conservative camp of astronomy, not easily won over by newfangled theories from "young Turks" such as Harlow Shapley. Curtis, slight, bespectacled, and balding, was a dedicated pipe smoker—infamous for starting small fires in his wastepaper basket when cleaning out his pipe.

The Lick Observatory had made an important contribution to the nebula debate even before Curtis's arrival. James Edward Keeler was director of the Lick Observatory from 1898 until his death just two years later. During his brief tenure at Lick, he had taken very long photographic exposures of portions of sky away from the

plane of the Milky Way with a new 36-inch telescope so as to detect extremely faint objects. He found tens of thousands of nebulae far too faint to have previously been seen, making earlier lists of
nebulae look positively puny. Typically, the nebulae had a spiral
structure that contained two intertwined trailing spiral arms, but
three or more arms were sometimes found.

A technique called spectroscopy proved of particular importance
in studying the nebulae. This technique separates light into its
component colors. In the 1860s a scientist named William Huggins
introduced spectroscopy to astronomy. The phenomena of colors,
in which a beam of white light passing through a glass prism produces the colors of the rainbow, had been demonstrated since ancient times. It was in the seventeenth century, however, that Isaac
Newton established, by a series of carefully thought-out experiments using the prism effect and logical reasoning, that white light
"is a confused aggregate of Rays endued with all sorts of colors." A
prism, by bending the different color components of white light by
varying amounts (red being bent least, violet most), separates a
beam of white light into a merging row of colors called a spectrum.
Since white light, passing through a slit and a prism, produces a
continuous range of colors, it is said to have a continuous spectrum.
In a continuous spectrum of white light, we have displayed before
us all the colors of the rainbow—red, merging with orange, merging with yellow, merging with green, and so forth through to violet. To take a contrasting example, a certain purple light source
might have only a red component and a blue component, so that
its spectrum would show just the two features, red and blue, on a
black background, making a so-called emission-line spectrum.
Thus if we scanned along the spectrum looking for a row of colors
we would at first see nothing but the black background, then a narrow strip of red, then black again before encountering the second narrow strip of blue, beyond which would lie more black
background.

Scientists found that spectroscopy, which involves the study of
features in a spectrum such as emission lines, could be used to infer the composition and nature of an object emitting light. Spectroscopy of the nebulae was to come into its own courtesy of an

astronomer called Vesto Melvin Slipher, who made many fundamental discoveries in the early years of the twentieth century. Born near Mulberry, Indiana, in 1875, Slipher was a pioneer of astronomical spectroscopy at the Lowell Observatory in Flagstaff, Arizona.

A wealthy Massachusetts doctor and amateur astronomer, Percival Lowell, had founded the Lowell Observatory in 1894. The formation of the observatory was in part based on a strange mistranslation from Italian to English. An Italian astronomer, Giovanni Schiaparelli, had noticed what appeared to be grooves on the planet Mars, which he described as "canali," meaning channels. When this work was translated into English in 1880, the word *canali* was reported as "canals"—implying structures that had been constructed by intelligent beings. This so-called evidence of intelligence on Mars sparked speculation about the nature of Martians—both in the scientific literature and in science fiction such as H. G. Wells's *War of the Worlds*. Percival Lowell was fascinated by the speculation about Martians and, with a vast family fortune behind him, built his observatory specifically to make detailed observations of Mars and the other planets. Unlike most patrons of science, Lowell made many of the observations himself, but he also recruited a small number of professional astronomers to help him.

Slipher joined the Lowell Observatory in 1901, when he was just twenty-five. He would stay there for fifty-three years. Lowell gave Slipher the task of studying nebulae, because one of the suggestions about spiral nebulae fashionable at that time was that they were swirling clouds of gas collapsing to become new planetary systems. By 1916 Slipher was the Lowell Observatory's acting director, although his appointment was not made substantive until 1926. Vesto Slipher used very long exposures of many tens of hours, over many nights, to collect sufficient light to get spectra with the most sensitive photographic plates then available. (The discovery of dry emulsions meant that astronomers could expose a single photographic plate over several successive nights.) In the era around 1910 he made the remarkable discovery that the spiral nebulae displayed spectra suggesting that they were made up of many different types of stars, so closely packed that they could not be resolved into individual stars by the telescopes then available. But Slipher made an even more startling observation. He found that although the principal spectral lines could be identified, they were almost always displaced toward the red end of the spectrum. Such a color shift may be explained in terms of the so-called Doppler effect.

Light is a form of wave—an electromagnetic wave. (In fact light is made up of discrete packets of energy called photons—but a continuous stream of photons can be readily visualized as a wave.) Any wave can be characterized by the distance between adjacent crests or troughs, a property known as wavelength. (Imagine ripples on a pond, where the wavelength is the distance between the ripples.) Light wavelengths are less than one millionth of a meter. And wavelength shift (meaning color shift in the case of visible light) is common to any wave motion, through the Doppler effect. The Doppler phenomenon relating to wavelength shift was first formulated mathematically in 1842 by Johann Christian Doppler, professor of mathematics at the Realschule in Prague. A familiar example of the Doppler effect is the sudden decrease in pitch noticed by a stationary observer of the siren of a passing police car. An approaching sound source bunches up the sound waves ahead of it, thus decreasing the sound wave's wavelength. A decrease in wavelength implies a corresponding increase in pitch. Similarly, when the sound source recedes, it stretches out the waves behind it so that the wavelength is increased and the pitch sounds low to a stationary observer. As it is with sound, so it is with light. The compression of wavelength from an approaching light source produces a blueshift, whereas a redshift is interpreted in terms of motion away from the observer. The degree of blueshift or redshift allows an estimate to be made of the speed of approach or recession of the source of light. Although a more subtle interpretation of the wavelength shift than the Doppler effect would be required, the measuring of redshift would prove later to be enormously powerful in estimating the size of the cosmos.

Slipher's redshifts were interpreted as motions of the spiral nebulae away from the solar system. But the speeds implied by the redshifts were so large as to be difficult to comprehend. One galaxy in particular, the so-called Sombrero Nebula (named after its shape), seemed to be moving at the quite staggering speed of 4 million kilometers per hour, dramatically faster than any other object ever observed to that time. Slipher reported his redshifts of fourteen spiral nebulae at the August 1914 meeting of the American Astronomical Society in Evanston, Illinois. Fourteen observations do not sound like very many, but it needs to be recalled just how difficult taking spectra of faint spiral nebulae was. There were hundreds of careful observing hours involved getting just this modest haul of results, with each spiral nebula requiring observations over many nights.

A member of the audience at the Evanston meeting was a young doctoral student from the University of Chicago called Edwin Hubble, who ten years later would use Slipher's results to redefine our understanding of the cosmos.

Following the announcement of his results, Slipher received a letter from an eminent Danish astronomer, Ejnar Hertzsprung:

> My hearty congratulations to your beautiful discovery of the great radial velocity of some spiral nebulae. It seems to me that with this discovery the great question of the spirals belonging to the system of the Milky Way or not is answered with great certainty to the end that they do not.

Slipher remained equally confident, writing in a paper read to the American Philosophical Society in April 1917:

> It has for a long time been suggested that the spiral nebulae are stellar systems seen at great distances. This is a so-called "island universe" theory which regards our stellar system the Milky Way as a great spiral nebula which we see from within. This theory, it seems to me, gains favor in the present observations.

But there were others who were not so sure! They would be most vociferous and influential in arguing that the Milky Way defined the whole cosmos. The controversy was getting more heated.

Heber Curtis was one of those who supported Slipher's results and interpretation. The redshift data certainly appeared to be clear evidence in support of the idea that the spiral nebulae were distinct island universes lying way beyond the bounds of the Milky Way, as far as Curtis was concerned. However, with the discovery of an increasing number of nebulae that displayed spectra confirming that they were merely clouds of hot gas rather than systems of billions of stars, many astronomers concluded that all forms of nebulae would in time be proved to be similar. They saw Slipher's results as a temporary embarrassment that would in time, hopefully, be resolved. A problem Slipher faced was that it was much easier for his rivals to analyze the spectra of bright gaseous nebulae that are close by than it was for him to detect the spectra of the very faint spiral nebulae. The island universe theory did not have many friends at this time, but its moment of glory was not far off.

The conflict between the different camps on the nature of the spiral nebulae has been likened to that within the Vatican in the sixteenth century between the Copernican and Ptolemaic worldviews—a head-on collision between two very different perspectives of the universe presented by those with equally strong but differing points of view. While this may be a rather extravagant claim as to the importance of the debate, the difference in worldviews was indeed marked. Shapley had produced evidence that the Milky Way was truly vast, so large in fact that one might be content that it represented the complete cosmos. In the Shapley worldview the nebulae would just have to be accepted as being gaseous clouds within the Milky Way (perhaps planetary systems in formation) or, at worst, small systems of stars appended like satellites to the Milky Way but of no particular significance on the grand cosmic scale—forget about the "island universe" nonsense. Curtis on the other hand was happier with a Milky Way of Kapteynian dimensions, allowing the solar system to lie near its heart. Curtis was convinced that, while some of the nebulae may indeed be clouds of hot gas within the Milky Way, the majority of spiral nebulae would eventually be shown to be distinct star systems at vast distances—Immanuel Kant's independent "island universes."

Curtis had initially accepted with considerable misgivings the invitation from Abbot and Hale to take part in the debate, but then, as he marshaled his arguments in the weeks before the meeting, his enthusiasm for the forthcoming encounter mounted. The more information he gathered to support his case, the more convinced he became that he was right. By contrast Shapley accepted the invitation with apparent enthusiasm; after all, George Ellery Hale was his boss, and he was flattered to be selected for such a prestigious event. However, Shapley would develop secret misgivings about the proposed encounter, which would grow as the days passed—despite the fact that, like Curtis, his faith in the correctness of his science remained solid. The reason for Shapley's misgivings was that he wished to avoid any prospect of public embarrassment ahead of a forthcoming prestigious appointment he hoped he might make his own.

In February 1919 the eminent director of the Harvard College Observatory, Charles Edward Pickering, had died. Pickering had been a giant of astronomy—the "benevolent dictator" who had built the Harvard College Observatory into an establishment respected around the world. Despite being only in his midthirties, Shapley rather fancied his chances as Pickering's successor. He wrote to his onetime teacher and mentor Henry Norris Russell and also to Hale, pressing his claim for Pickering's crown. Russell believed that Shapley was a gifted astronomer who would be wasted having to concentrate his efforts on the routine administration that commanded so much of the time of any observatory director. He was not encouraging in his reply to Shapley:

> To tell the naked truth, I would be very glad to see you in a good position at Harvard, free from executive cares. . . . But I would not recommend you for the Pickering place; and I believe that you would make the mistake of your life if you tried to fill it.

And Russell confided his misgivings to Hale in a somewhat enigmatic note:

> Shapley would not suffer if he pondered the old fairy tale about the man who got all sorts of good things from a magic fish whose life he had saved—until his wife wanted to be Pope.

The implication is that Russell saw himself as the "magic fish" that had granted Shapley many wishes—but he was not prepared to grant this one.

In early 1920 a Harvard official visited Shapley to discuss his possible appointment to the Harvard College Observatory. But, unbeknown to Shapley, Harvard at this stage (perhaps encouraged by Russell) had in mind the position of an astronomical assistant to the director rather than the actual director's post that Shapley coveted. Convinced that he was in line for the director's post, Shapley viewed with some dread the prospect of a high-profile and much publicized encounter with Curtis—especially once he heard that Harvard would be sending some of its senior people to the meeting to see the debaters in action. Curtis was well known as a skillful public speaker, able to draw on the eloquent expository style he had perfected in the classroom. Shapley was not confident that he could match Curtis's debating skills, even if he remained supremely confident in his scientific case. And if he suffered public humiliation at the hands of Curtis, Shapley feared that this would diminish the likelihood of his getting the Harvard directorship. Shapley did try, unsuccessfully, to get someone other than Curtis appointed as his adversary. And he argued that rather than a formal debate, perhaps two talks on the same subject, albeit from differing perspectives, might be a more worthwhile format. But Curtis was now warming to the idea of a confrontation of opposite worldviews. Shapley did not find a letter from Curtis reassuring. Displaying his Irish heritage, Curtis wrote to him:

> I agree with you that it should not be made a formal "debate," but I am sure that we could be just as good friends if we did go at each other "hammer and tongs." A good friendly "scrap" is an excellent thing once in a while; sort of clears up the atmosphere. It might be far more interesting both for us and our jury, to shake hands, metaphorically speaking, at the beginning and conclusion of our talks, but use our shillelaghs in the interim to the best of our ability.

A verbal fisticuffs against a skilled orator was most certainly not what Shapley had in mind, especially if Harvard was planning to

send senior representatives to witness the event as part of its assess-
ment of Shapley's suitability, as he thought, to take over the Harvard College Observatory. Curtis had sent a copy of his letter to
Shapley on to Hale. Shapley did manage to persuade Hale to move
away from his original idea for a debate and, fearing that he would
not compare well in terms of lengthy presentation, asked that the
talks be shortened from the proposed forty-five minutes to thirty-
five minutes each.

Letters were fired back and forth between Hale, Shapley, and
Curtis. Curtis was furious that Shapley wanted the talks shortened:
"we could hardly get warmed up in 35 minutes!" They compro-
mised on forty minutes. And what about the format? In a letter to
Curtis, Hale insisted on having the final word:

> I do not think that the discussion should be called a "debate,"
> or that Shapley, who is perfectly willing to speak first, should
> have time allotted him for a "rebuttal." If you or he wishes to
> answer points made by the other, you can do so in the gen-
> eral discussion. . . . Each speaker should be manifestly a seeker
> after truth, willing to point out the weak places in his argu-
> ment and the need for more results.

A format for the event was eventually agreed. Each debater
would be given forty minutes for his presentation and also given a
single opportunity to refute proposals from the other. All other
contributions would come from the audience. Neither Curtis nor
Shapley had got what he had wanted; each settled for a compro-
mise that would not give too obvious an advantage to his adversary.

Shapley decided to bring in a strong armory to support him during the meeting. He connived with Hale for Russell to be called on
to provide the first comment from the floor after he and Curtis had
made their presentations. Of course Russell was his close friend and
mentor—and was a solid believer that the Milky Way was not
confined to Kapteynian dimensions. Shapley had already headed
off the idea of a face-to-face argument and counterargument in traditional debating style, fearing Curtis's superior debating skills.
Nevertheless he needed someone to challenge Curtis if the style of

encounter he had insisted on would not enable him to do this directly. Russell was so widely respected that Shapley felt that a word of support from such a powerful ally would prove invaluable. He wrote in advance to Russell:

> I lead off. . . . Then Curtis presents his views, and then follows general discussion. Mr Hale is anxious that you lead that discussion in whatever way you see fit, and I believe he plans to ask the presiding officer to call upon you as a starter. Curtis . . . will show (?) that my distances are some ten times too big. Now that ten times, as Mr Hale realizes, is as bad in your hypotheses as in mine; it is a violation of nearly all recent astrophysical theory. So unless Curtis bowls us over with the only true truth in these celestial matters, you will be interested in this general assault from the self-styled conservatives. . . . [Some people] at Lick and Mount Wilson seem to regard that coming discussion as a crisis for the newer astrophysical theories. . . . But crisis or not, I am requested to talk to the general public of non-scientists that may happen to drop in. Consequently, whatever answer must be made to Curtis and his school must be made in the discussion. I write you this because you may be interested in knowing what the situation is, and so that you may be ready to defend your own views if they are imposed upon by either of us.

This final sentence is disingenuous, since it could only be Curtis who might "impose" on Russell's views (Shapley knowing that they coincided so closely with his own). In the event, Russell made such a strong contribution from the floor in support of Shapley that few present (including the visitors from Harvard) could have doubted that there had been a degree of connivance between the two friends prior to the event.

Conviction and self-confidence were present in abundance on both sides of the arguments. Approaching the outset of the Washington meeting, Curtis and Shapley had divergent views, not only on the scientific issues, but also, it seems, on what the subject of the debate was supposed to be. Shapley had decided to limit his presentation to the size of the Milky Way, making only very brief reference to the spiral nebulae. By contrast Curtis had decided that the nature of the spiral nebulae should be considered and would represent the major part of his presentation. The difference in emphasis is surprising, in view of the extent of prior correspondence. It is difficult to escape the conclusion that Shapley sought to avoid a confrontation on the nature of the spiral nebulae, so as to diminish the opportunity for Curtis to expose his arguments to ridicule. He positioned himself firmly on the familiar ground of measuring the size of the Milky Way.

The Great Debate took place in the main auditorium of the U.S. National Museum, now the Natural History Museum. The auditorium is still in use today and is now known as the Baird Auditorium. The audience was a mixture of professional scientists and possibly an equal number of members of the general public. One record of the event suggests that Albert Einstein was in the audience. The two speakers needed to depend on the natural projection of their voices, since no amplification was provided in such auditoriums in 1920. For an experienced teacher, articulate, confident, and self-assured this would not present a problem; Curtis was in his element. Shapley was in a less familiar environment.

Curtis's presentation was dominated by his insistence that the spiral nebulae were distant massive island universes (galaxies), while Shapley spent just a few minutes arguing that the spirals were small and nearby, instead spending most of his time trying to persuade the audience that the Milky Way was very much larger than previously believed.

The difference in the style of presentation of the two speakers was very marked. Shapley presented his talk in general terms, sensing (in a way that Curtis failed to) the diverse range of scientific interests of the audience. He therefore had taken great pains to prepare a talk that might be understood by someone with only a most modest understanding of the rudiments of astronomy. The nineteen-page manuscript for his talk still exists. It is not until page 6 that he reaches the definition of a light-year! By contrast Curtis presented highly technical arguments, stretching the understanding of even those few specialists in the audience. He had decided to present his talk through a series of typewritten slides packed with detailed information likely to be understood only by those fully immersed in the subject of discussion. Copies of the slides (if not his notes) survive, and they testify to the relative complexity of Curtis's address. Since the event was open to the general public, the level of Curtis's talk was not entirely appropriate. In later correspondence the debaters acknowledged the difference in style. Hale had suggested that Shapley and Curtis should publish the arguments presented in their talks. Shapley later wrote to Curtis:

[Hale favors publication]—even if the papers are long, providing the material is suitable in being not too popular (like mine?) or too tabular or technical (like yours?)

Curtis acknowledged that his presentation might indeed have been too specialized:

Yes, I guess mine was too technical. I thought yours would be along the same line, but you surprised me by making it far more general in character than I had expected. I had some thought of changing entire character of my presentation about five minutes before close of your part, but decided at last minute to go ahead with program as planned.

The final sentence implies that Curtis felt it was Shapley who had departed from an agreed format of expert evidence. Indeed, from the nature of the correspondence before the event, one can understand Curtis's sense of frustration that Shapley, by trying to popularize his presentation, had been rather too liberal in his interpretation of the rules of engagement.

The evidence Curtis had accumulated in support of his assertion that the spiral nebulae were separate island universes at vast distance from the Milky Way was persuasive. The basis of his case was as follows. There was a huge spread of angular sizes seen for the nebulae, ranging from some with such significant angular dimension that details of the spiral structure were obvious with even a small telescope, down to others that were so tiny that their spiral structure could barely be discerned with the most powerful telescopes then available. Such a range of angular sizes was surely indicative of a vast range of distances, with the large ones being relatively nearby and the tiny ones being at extreme distance. Besides, as more and more powerful telescopes were being built, more and more spiral nebulae were being found. This was not the case for stars, suggesting that smaller telescopes had seen to the edge of the Milky Way and the stars it contained, and that the new, larger telescopes were finding the faint island universes lying far beyond and for which no comparable outer boundary had yet been detected. And while stars had been found to be concentrated along the plane of the flattened disk of the Milky Way, spiral nebulae were being found scattered around the celestial sphere in what Curtis described as "an apparent abhorrence of our galaxy of stars."

Curtis was impressed by the fact that the spectra of spiral nebulae were "practically the same as that given by a star cluster," albeit that the spectra of spiral galaxies tended to be redshifted to varying degrees. And since the redshifted spectra suggested that the spirals were traveling at considerable velocity, yet there was no discernible proper motion of the nebulae across the sky, then one had to assume that they were at vast distance. The large velocities implied by the redshifts convinced Curtis that the idea that spiral nebulae were rotating clouds of gas collapsing to become new stars was untenable; surely such clouds would be traveling through space with comparable velocities to the stars they would become—not twenty times faster!

There was another key argument that Curtis believed supported the island universe hypothesis. Although individual stars could not be discerned in distant spiral nebulae, it should be possible to detect stars that brightened suddenly and spectacularly. Such "new stars," or novae, had been recorded from very ancient times with the naked eye (especially in China), and about thirty had been studied in detail since the advent of the telescopic era. As Curtis described them, "All [novae] have shown the same general history, suddenly increasing in light ten thousand-fold or more, and then gradually, but still relatively rapidly, sinking into obscurity again. They are a very interesting class, nor has astronomy as yet been able to give any universally accepted explanation of these anomalous objects." Curtis believed that if novae were a regular part of the Milky Way, then they should also be a regular part of spiral nebulae if the spiral nebulae were island universes. And sure enough, in many spiral nebulae new star events had been recorded that looked suspiciously like novae. The only difference was that these events in spiral nebulae were very much fainter than those seen in the Milky Way—exactly what one would expect if they were at vast distance, lending further credence to the island universe hypothesis. There was, however, one nova that had occurred in a famous nebula in the constellation Andromeda in 1885 that apparently presented a counter view. It had been so bright that it practically obliterated the light from the nebula, and if it had been a nova then this would imply that the Andromeda Nebula must lie relatively nearby, within the Milky Way. But if the Andromeda Nebula was an independent "island universe" at vast distance, then the 1885 new star must have been as bright as a billion Suns—intrinsically far more brilliant than any known nova and counter to everything astronomers then knew about the cosmos. (We now know that such "super" brilliant stellar outbursts do occur, as the supernovae that will be discussed later. But the notion of such "super" novae was not appreciated in 1920.) Curtis wanted the 1885 outburst, and another similarly spectacular outburst in a different spiral nebula ten years later, to be seen as special cases of spectacularly brilliant novae way beyond the norm. However, for many the spectacular brightness of the 1885 outburst meant that the island universe hypothesis had to be discounted. A contemporary report noted that, for the spiral nebulae,

> It would be eminently rash to conclude that they are really aggregations of Sun-like bodies. The improbability of such an inference has been greatly enhanced by the occurrence, at an interval of a quarter of a century, of stellar outbursts in two of them. For it is practically certain that, however distant the nebulae, the stars were equally remote; hence, if the constituent particles of the former be suns, the incomparably vaster orbs, by which their feeble light was well-nigh obliterated

must have been a scale of magnitude such as the imagination recoils from contemplating.

Astronomers would have to learn to contemplate the totally unexpected, without recoiling!

To Curtis the evidence he had accumulated ahead of the Washington meeting was indisputable. One could settle for the Milky Way having Kapteynian dimensions (with the Sun near its center)—just one galaxy among a cosmos generously populated with distinct island universes. He was indeed well prepared for the debate. No wonder he was confident about his position and was relaxed about having been asked to follow Shapley's presentation.

What possible defense could Shapley have offered as counterargument to such a barrage of persuasive evidence for island universes? Actually Shapley refused to be drawn into the details of the spiral nebulae. Instead he changed the whole basis of the confrontation, so as to avoid Curtis's arguments. Shapley wished to concentrate on the irrefutable evidence he had produced that the Milky Way was spectacularly larger than the estimates of Kapteyn and that the solar system lay in its outer reaches. He only briefly mentioned the spiral nebulae at the end of the lecture, so to a certain extent the two "debaters" were talking past each other.

The reason Shapley was so convinced that the spiral nebulae were nearby was because of the work of a close colleague from Mount Wilson, a Dutch astronomer called Adriaan van Maanen. Van Maanen had used a device known as a blink comparator to compare photographic plates taken at different times—say months or years apart—to try to detect subtle movements of objects in the heavens. Using the blink comparator van Maanen had convinced himself, and Shapley, that he could detect rotational movement of examples of spiral nebulae viewed face-on. The detection of such subtle rotations would be possible only if the spiral nebulae were relatively nearby, that is, if they were lying within the Milky Way. For a face-on spiral nebula called M101 (object 101 in a catalogue of nebulae prepared by French astronomer Charles Messier), also known as the Pinwheel Nebula since it looks like a rotating firework with sparks flying off, van Maanen claimed that he could detect a minute rotation of 0.02 seconds of arc each year. This is a very small angular displacement indeed, but it would nevertheless be of great significance if real in determining the proximity of the spiral nebulae. Van Maanen claimed to have detected similar subtle rotations in six other spirals. If the spiral nebulae were enormous structures at vast distance, then any discernible rotation could not be expected in times shorter than tens of thousand of years. The angular rotation van Maanen had measured would demand that the spiral nebulae were rotating at the speed of light if they genuinely lay at vast distance—a quite ridiculous proposition.

We now know that van Maanen was wrong in claiming to have observed rotation of the spiral nebulae. Subsequent investigation has shown that neither the photographic plates nor the blink comparator were at fault, so just how van Maanen's errors arose remains one of astronomy's more intriguing mysteries. However, it was many years after the Great Debate before the error was revealed and Shapley would be forced to reconsider his opposition to island universes.

At the Washington meeting, Shapley tried to diminish the strength of Curtis's arguments even before Curtis had made his presentation by suggesting that any definitive conclusions about the nature of the spiral nebulae would be premature when they were now being subjected to such intense investigation by many eminent astronomers—implying that more data were needed before Curtis's case could even be seriously considered. He said in his presentation:

> On one point [Curtis and I] agree, or at least we should agree, and that is that we know relatively so little concerning the spiral nebulae and we are soon going to know relatively much because of the increasing activity in the nebular field, that it is professionally and scientifically unwise to take any very positive view in the matter just now.

Curtis must have been furious that Shapley had tried to negate the whole of his talk with such a dismissive judgment before he had had the chance even to present it! It is true that Hale had written that each debater should be "willing to point out the weak places in his argument and the need for more results." But Shapley had unreasonably chosen to point out the need for more results from his opponent, not himself. This looked like intellectual skulduggery. However, Curtis was not above aiming a blow beneath the belt himself. In referring to van Maanen's observations, used by Shapley to argue that the spiral nebulae lay nearby, Curtis retorted:

> There are some observations that are now worth a damn— and others that are not worth a damn. In my opinion two damns are not better than one damn.

The laughter from the audience hinted that Curtis was winning their affection, if not their understanding, with his spontaneous and enthusiastic style; and Shapley must have felt some embarrassment in front of the Harvard visitors.

Reflecting on the event later, Curtis felt he had won the Great Debate, even though he had used a level of address inaccessible to many in the audience. He certainly had friends who had been present who agreed with him that he was the winner. Writing to his family, Curtis noted: "Debate went off fine in Washington, and I have been assured that I came out considerably in front." But Shapley had friends who assured him that *he* had in fact won. Russell was not so sure and patronizingly suggested that his friend should offer some lecture courses to hone his presentational skills. In later life Shapley was honest enough to record: "Now I would know how to dodge things a little better. As I remember it, I read my paper and Curtis presented his paper, probably not reading much since he was an articulate person and was not scared."

However, new results just four years later would demonstrate that while Shapley was correct in dislodging the Sun from the center of the Milky Way, he was wrong in challenging the fact that the spiral nebulae might be island universes. It was Curtis who had been right all along in arguing that the spiral nebulae were distinct galaxies lying far beyond the bounds of the Milky Way (although Curtis had also made mistakes in limiting the size of the Milky Way and retaining the Sun at its center). Since they were both right in part, and both wrong on key issues, perhaps history would be wise to declare the Great Debate of 1920 a draw.

The Shapley-Curtis debate represented a clear break between a nineteenth-century worldview that limited the size of the cosmos to the Milky Way and the eventual twentieth-century acceptance of a vast cosmos made up of billions of galaxies separated by enormous distances. The debate's true significance was not understood at the time, and it was not widely reported in the press. Even the speakers did not appreciate how the debate catalyzed a new understanding of the cosmos. The Great Debate is now commemorated regularly by the National Academy of Sciences, as contemporary astronomers pit their wits against one another on topics of current controversy in new "great debates."

And where did Hale, Slipher, Shapley, and Curtis end up? Suffering from overwork, and after recurrent episodes of depression, Hale resigned as director of the Mount Wilson Observatory in 1923 and withdrew from active involvement in scientific matters while still arguably in the prime of his career. He died on February 21, 1938, in Pasadena, California. He is remembered as perhaps the greatest of all telescope builders.

Vesto Slipher served as the director of the Lowell Observatory until 1956. His best-known work, other than his spiral nebulae redshift observations, was leading the team that discovered the planet Pluto in 1930. Sadly Slipher's manifold contributions to astronomy never received the attention they deserved; his colleague Clyde Tombaugh received all the publicity for the discovery of the planet Pluto, although it was Slipher's inspired leadership that had really made it possible. He died at Flagstaff on November 9, 1969.

Shortly after the 1920 debate Curtis was appointed by the University of Pittsburgh to the directorship of the Allegheny Observatory, and later, in 1931, he was appointed to the directorship of the University of Michigan observatories. A severe thyroid disease sadly plagued his final years. He died at the observatory residence at Ann Arbor on January 9, 1942. His work at Lick was the pinnacle of his research career, but he proved himself to be an able administrator in his later observatory directorship posts. He does not figure as prominently in the history books of astronomy as one might expect for someone who helped change the vision of the cosmos in such a fundamental way. It is also sobering to reflect that during the lifetime of the son of a Union veteran, not only had our understanding of the nature of the cosmos been revolutionized, but the United States had been transformed from a "nation divided" to a global military, scientific, and technological superpower.

After the Great Debate Harlow Shapley did indeed secure the coveted directorship of the Harvard College Observatory, serving with immense distinction for thirty-one years and making monumental contributions to astronomy. In 1952 he was made director emeritus of the Harvard College Observatory and Paine Professor of Astronomy. He died at Boulder, Colorado, October 20, 1972. An obituary placed his greatness in locating the center of the Milky Way and the Sun's location far from it:

> Harlow Shapley was an outstanding man of his time—astronomer, educator, orator, as well as man of affairs. Some of his gifts displayed prominently in the course of his life, may gradually fall in oblivion as those of us who knew him in his prime may no longer be here to remember, and dust may settle on some of his work, or on many honours bestowed upon him by his contemporaries. But one title to fame will never tarnish—Shapley's discovery of the center of our Galaxy, and our position within it.

The great debaters had passed into history, but their Great Debate and its aftereffects reshaped human understanding of the cosmos.

Chapter 4

SEEING RED

The discoveries leading up to the Great Debate suggested that both the Milky Way and the entire cosmos might be bigger than anyone had anticipated. Parallax measurements were already producing some stellar distances (though admittedly in rather limited numbers). In the late nineteenth century, astronomers—many of them women—would find new methods that complemented parallax and extended scientists' reach to even greater distances. When astronomers combined these tools with Albert Einstein's new theories of relativity, they reached a shocking conclusion: the universe was not only big, it was getting bigger. The cosmos was expanding.

The opportunities for women in any profession, let alone one as historically male-dominated as astronomy, were very limited in the first decades of the twentieth century when compared with today (although even today evidence of gender inequality remains in the world of science). There was, however, one woman who was to rise above the conventional prejudices of the era, and her influence would prove to be truly profound. She was to provide the key that would unlock the mystery of the spiral nebulae and eventually resolve the Great Debate of 1920. Yet she achieved this against great odds.

Henrietta Swan Leavitt was born on July 4, 1868, in Lancaster, Massachusetts. It was an appropriate birth date for a true American patriot who would bring credit to her country and to her profession. She was the daughter of a Congregational minister. Following a strict upbringing and education at Oberlin College, she attended the Society for Collegiate Instruction of Women—later to become Radcliffe College, the women's college affiliated with Harvard University. (In 1999 Radcliffe formally merged with Harvard University, becoming the Radcliffe Institute for Advanced Study.) As a senior in 1892, Leavitt was introduced to astronomy. She was fascinated by it, and after graduation she enrolled in a course to study the subject full-time. Tragically Henrietta Leavitt was suddenly struck down by a serious illness, and she was forced to spend over two years at home recovering. Her illness left her profoundly deaf. She had not forgotten her love of astronomy, however, and when she felt fit enough she put forward her name in 1895 as a volunteer worker at the Harvard College Observatory.

The director of the Harvard College Observatory at the time was the inimitable Charles Edward Pickering, who had been appointed director in 1876 at the age of thirty. (It was Pickering's death in 1919 that would lead to Harlow Shapley's move to Harvard following the 1920 Great Debate.) Pickering took on many women as unpaid volunteers and a few in paid positions, as "computers"—those who undertook the laborious scanning and measuring of photographic plates and the resulting calculations of the positions and the brightness of stars. This style of employing women "computers" was part of an established pattern in astronomy, for example at the Royal Observatory at Greenwich and the Naval Observatory in Washington, D.C. (The women who worked with the computers to record their results were called "recorders." The women computers were paid little enough, but the recorders were paid a trifling "50 cents in the dollar" of the derisory rate offered the computers!) Leavitt worked with another woman computer called Annie Jump Cannon, five years her senior. Cannon coincidentally was also deaf, and was an inspiration to Leavitt in living with her deafness while pursuing a career in astronomy.

Annie Jump Cannon was an expert in the spectral classification of stars. She had been born in Dover, Delaware, on December 11, 1863, the daughter of state senator Wilson Cannon and his second wife, Mary Jump. Like Leavitt, Cannon was educated at Radcliffe College, before being recruited by Pickering as a paid computer.

Her career's output was staggering—including the nine volumes of the monumental *Henry Draper Catalogue* of almost a quarter million stars (so-called since the work was sponsored in the memory of Henry Draper, a wealthy physician and amateur astronomer, by his widow). The catalogue is still accepted as an international classic. The later *Henry Draper Extension* brought the number of stars Cannon classified and catalogued in her lifetime to 330,000—a truly Herculean task. She was the first woman to receive an honorary doctorate from the University of Oxford. Cannon's classification of stars, according to the characteristics of their spectra and color, would prove invaluable in later studies of galaxies based on the nature of their brightest stars.

From a star's spectrum we get a clue to the elements emitting light, and to their relative abundance. Stars are assigned to several possible classes, labeled O, B, A, F, G, K, M, R, N (remembered by the mnemonic "Oh Be A Fine Girl, Kiss Me Right Now"). This sequence is a temperature sequence, with O stars being the hottest and the later classes being cooler.

Cannon took great pride in her work and once wrote:

Classifying the stars has helped materially in all studies of the structure of the universe. No greater problem is presented to the human mind. Teaching man his relatively small sphere in the creation also encourages him by its lessons of the unity of Nature and shows him that his power of comprehension allies him with the great intelligence over-reaching all.

Henrietta Leavitt worked on what is called photographic photometry, the measuring of the brightness of a star (the so-called visual magnitude) from its size on a photographic image taken through a telescope. This was painstaking and laborious work, requiring a keen eye and considerable patience. Leavitt was to make the art of photographic photometry her own in the first decade of the twentieth century.

Pickering gave Leavitt the job of looking for interesting stars on photographic plates obtained from Harvard's southern telescope in Peru. (Pickering's brother, William, operated the telescope in

Peru.) The southern skies had not been studied as fully as the northern skies, and they brought forward a rich harvest of astronomical results. Leavitt took a particular interest in stars whose brightness varied with time, often in a regular fashion such that a star brightens then fades, brightens then fades, every few days to tens of days. The time between brightness maxima for such a regularly varying star is called its period. It was in 1784 that the first star to display such periodic variation in brightness had been discovered, by the Englishman John Goodriche. It was a star called delta Cephei, the fourth-brightest star in the constellation Cepheus. (The brightest star in a constellation was signified by alpha, the second-brightest by beta, and so forth.) The regular interval over which delta Cephei varied was found to be 5.37 days. As other regularly varying stars were found, they were given the collective name Cepheid variables (or simply Cepheids), after the first in their class.

It is now known that the varying brightness of Cepheids results from an actual physical pulsation—the periodic subtle increasing and decreasing in the size of a giant star, with attendant variation of surface temperature and thence brightness. Interestingly Goodriche, the discoverer of delta Cephei, was deaf like Leavitt, his intellectual descendant who would make the Cepheid variables famous.

Through her meticulous study of photographic plates from Peru, comparing the brightness of a myriad of stars, night on night, week on week, month on month, year on year, Henrietta Leavitt discovered 2,400 variable stars—almost half the total number of variable stars known in her lifetime. The Harvard computers were nothing if not dedicated and hardworking, and faced laborious tasks with remarkable commitment and good humor. Photographs of the twenty or so computers and recorders taken about 1910 display groups of smiling, trimly dressed ladies giving every impression of a happy (albeit shamefully exploited) team.

When it came to finding variable stars, Leavitt had no peer. Even today, with all the power of modern instruments and automated measuring techniques, Leavitt's catalogue of variable stars represents almost 10 percent of those with catalogued behavior. Hers was a labor of heroic magnitude, bearing in mind the technology available in her day and the male-dominated environment in which she worked (not to mention her serious deafness).

Leavitt made her most profound contribution to the understanding of the true scale of the cosmos through careful observation of Cepheid variables. The Large and Small Magellanic Clouds appear as diffuse nebulous bands in the southern skies. They were named after the Portuguese navigator Ferdinand Magellan, whose crew sighted them during the first circumnavigation of the world in the early sixteenth century and reported their existence to European scholars (although the Magellanic Clouds are recounted in the mythologies of Southern Hemisphere peoples from ancient times).

In Leavitt's time the Magellanic Clouds were thought to be clusters of stars lying on the outer periphery of the Milky Way (rather than as we know them today, as distinct and distant satellite galaxies to the Milky Way); or perhaps, some suggested, they were fragments of the Milky Way that had broken away to become relatively nearby satellites to the main body of the Milky Way. Of Leavitt's 2,400 variable stars, 1,777 were in the Magellanic Clouds, with the rest lying within the Milky Way.

Leavitt concluded that it would be worth looking at the Cepheids in the Magellanic Clouds to see whether there was a relationship between the period of any individual Cepheid and its average maximum brightness. She found that the brighter the Cepheid, the longer its period. But since all Cepheids in the Magellanic Clouds can be assumed to be at roughly the same distance from Earth (the size of this conglomerate of stars being small when compared with the likely distance of the Magellanic Clouds from Earth), then the relationship of the period would actually be with their intrinsic brightness (known as their absolute brightness) and not just their apparent brightness (which would depend on distance). She published her results, giving the so-called period-luminosity relationship for twenty-five Cepheids in the Small Magellanic Cloud, in 1912.

In 1913 the Danish astronomer Ejnar Hertzsprung accurately estimated the distances of a few Cepheids lying within the Milky

Way by parallax measurements, thus relating their absolute brightness to period in these few cases. With the parallax calibration provided by Hertzsprung, the distance to any Cepheid could be estimated from Leavitt's period-luminosity relationship. Astronomers had been provided with their next yardstick, after parallax, for measuring the size of the universe.

A simple thought experiment demonstrates how the period-luminosity (P-L) relationship is used for providing distance estimates. Imagine you are driving along a straight road at night and see a faint light. You have to make a judgment as to whether this is the inherently faint light of a nearby approaching bicycle or perhaps the inherently bright headlight of a Harley Davidson that appears to be faint merely because it is still some distance off. You would look for further evidence—perhaps seeking out the silhouette of a cyclist in your headlights, or some estimate of increasing brightness that would be evidence of the speed of an approaching Harley Davidson. But just suppose the manufacturers of lights for bikes (of all kinds) had decided to endow them with a "period-luminosity" relationship, so that inherently fainter lights for cycles had been designed to flick off, then rapidly on again, every three seconds—while the intrinsically very bright lights of Harleys had been designed to flick off, then rapidly on again, every ten seconds. So now when you see a faint light you can easily check whether it is flicking off briefly every few seconds, in which case it is inherently faint and nearby, or is flicking off briefly on a much longer period, in which case it is intrinsically very bright but is appearing faint because it is very distant. Such a thought experiment demonstrates an imaginary "period-luminosity relationship" for bikes. However, nature has endowed Cepheids with a real period-luminosity relationship. Thus Cepheids can be used as the much sought-after "standard candles" first postulated by Herschel: objects of known intrinsic brightness that can be used to establish distance. By observing a Cepheid period, one can determine how bright it actually is (its intrinsic brightness); and by comparing its intrinsic brightness with the observed brightness, one can then determine its distance.

Perhaps it is not surprising that Charles Edward Pickering

presented himself as the "author" of the paper on the period-luminosity relationship. As director of the observatory he clearly thought it was right for him to make the announcement. However, everyone knew that it was really Henrietta Leavitt's work.

Despite Leavitt's spectacular results with Cepheids, Pickering insisted that she must now move to a different project. Hence others continued her pioneering work. She did eventually become head of the Photographic Photometry Department at the Harvard College Observatory. In this role she developed a standard of photographic measurements of star brightness that was accepted by the International Committee on Photographic Measurements in 1913 and called the Harvard Standard. Perhaps it could more justly have been memorialized as the "Leavitt Standard."

Because of the prejudices of the day, women working in astronomy were expected to play largely subordinate roles, and Leavitt never really had the chance to fully exploit her intellect, follow her own inclinations, or pursue new avenues in research in the way her male colleagues were able to do. It is intriguing to speculate as to whether, if Pickering had left her free to pursue the consequences of her P-L relationship for Cepheids, she herself might have solved the mysteries of the cosmic distance scale and determined the distances to the spiral nebulae well ahead of the Great Debate. One suspects she almost certainly would have risen to this challenge.

Henrietta Leavitt never complained about her lot. But a male colleague at Harvard freely acknowledged that at this time "she possessed the best mind at the Observatory." Contemporaries recognized her as one of the most brilliant women of her generation at Harvard. And her epoch-defining role in helping reveal cosmic distances meant that she was given serious consideration to be a recipient of a Nobel Prize (she was nominated by the Swedish mathematician Mittag Leffler). The Swedish Academy of Sciences was struggling at the time over the issue of whether astronomy could be included within the physics prize.

Sadly, Henrietta Leavitt died of cancer in 1921, when just fifty-three years old. With her deafness, her struggles to gain recognition in a male-dominated profession, her failure to be awarded a Nobel Prize despite being given serious consideration, and her premature death, destiny had dealt Henrietta Swan Leavitt a cruel hand. She did not live to see how Edwin Hubble, in 1924, would use her research to finally resolve the issue of the nature of the spiral nebulae.

Henrietta Leavitt's name has been given to a crater on the Moon, to honor deaf men and women who have made magnificent contributions to science. This is a modest, albeit unusual, memorial to a woman who made such a major contribution to astronomy. A grander and more eloquent gesture should have been tendered by the astronomical community to such a key figure in the history of their subject. Annie Jump Cannon did receive many honors during her lifetime and also had a lunar crater named after her. But the prejudices of the time were such that her position at Harvard was not made permanent until 1938, just two years before she retired. She died only one year after retirement, on April 13, 1941.

If Henrietta Leavitt had provided the key to determine the size of the cosmos, then it was Edwin Powell Hubble who inserted it in the lock and provided the observations that allowed it to be turned. Edwin Hubble had been born in Marshfield, Missouri, on November 20, 1889, when his parents were visiting his grandparents. His father, John Powell Hubble, a failed lawyer, worked for an insurance company in Chicago, and the family settled in Wheaton. Edwin Hubble was a keen reader as a child and loved the novels of Jules Verne—perhaps inspiring the young Edwin to consider exploring the unknown. His interest in astronomy was inspired by a grandfather who owned a small telescope, and a letter he wrote to his grandfather about the planet Mars so pleased the old man that he had it printed in the Springfield newspaper, making it Edwin Hubble's first scientific publication. Like many youngsters, he earned his pocket money by delivering newspapers. He attended Wheaton High School. He was exceptional at sports, excelling in athletics, boxing, and football. The high school gave him a scholarship to the University of Chicago, although through an administrative error the single scholarship for the year had also been promised to another boy, so it had to be shared at half the value each. He made up the difference by being taken on as a laboratory assistant to the Nobel laureate Robert Millikan (the man who determined the electrical charge of an electron) and also by taking on tutoring and summer work. At the university he excelled in track and field, basketball, and boxing. (Fight promoters supposedly encouraged him to turn pro, although it is difficult to test the reliability of this story, since when he was older Hubble had the tendency to embellish stories of his younger years and to make somewhat exaggerated claims about his sporting prowess. But there can be no doubting that Hubble, at six feet two inches and close to two hundred pounds, was blessed with a degree of youthful athleticism.) He graduated in 1910 with a degree in mathematics and astronomy.

Hubble was charismatic, and he cultivated a "larger than life" aura that served him well throughout his life. With good looks, a keen intellect, and noteworthy sporting ability, his future was assured. But it was not certain that this future would be in astronomy, since parental encouragement pointed in the direction of the law. In 1912 he was awarded a prestigious Rhodes scholarship to study Roman and English law at Queens College of Oxford University in England. He loved his time at Oxford, with its medieval quaintness, the traditions, and the English eccentricities. This was the Oxford of *Brideshead Revisited*: a university dominated by male-only colleges, populated mainly by privileged young men from wealthy families who were nurtured through the English elite public school system before "finishing" at Oxford prepared them for a life of high accomplishment in the civil service, the military, the diplomatic service, the law, or merchant banking. Within this established elite, Rhodes scholarships carried special weight. During his time at Oxford Hubble acquired a number of affectations and an emphasized English diction that remained with him throughout his life. (His affected Englishness was a cause of some amusement to fellow astronomers in later years.) Popular with other students at Queens, he seemed to have a particular attraction for the young ladies. He continued his sporting interests and took part in one of the first basketball matches in Britain—although his sporting achievements at Oxford never quite hit the peaks that he would later claim.

Hubble gained an Oxford law degree, then returned to the

United States in 1913 and settled in Louisville, Kentucky, where his family had moved. (Years later he was made an Honorary Fellow of Queens College—in recognition of his contributions to astronomy, it has to be said, rather than his contributions to law.) Honoring a promise to his dying father, Edwin Hubble tried to move into law practice but had to settle for high school teaching. But astronomy was in his blood, and in 1914 he moved to the University of Chicago's Yerkes Observatory (the observatory founded by George Ellery Hale) to take up a research post. He reported at the time:

> I chucked the law for astronomy, and I knew that even if I was a second-rate or third-rate astronomer, it was astronomy that mattered.

In truth, he had "chucked" schoolteaching for astronomy, but the idea of walking away from the law profession for astronomy appeared to fit better the mythology Hubble sought to develop.

Hubble started studying for a Ph.D. at Chicago, working on nebulae. His work on the classification of nebulae did not make a major impact at this time but was sufficient to gain him his doctorate. There was still great uncertainty surrounding the nature of the nebulae, with conventional wisdom at this time still favoring their being gaseous clouds lying within the Milky Way. The grainy photographic images possible for such faint objects failed to resolve any stars that might be integral components of the nebulae, other than in exceptional cases.

While finishing his doctorate early in 1917, Hubble was invited by George Ellery Hale to join the staff of the Mount Wilson Observatory. It was a great honor, and a wonderful opportunity. However, fate—in the form of the First World War—intervened. After sitting up all evening to finish his doctoral dissertation, and taking his oral examination, he headed for the army enlisting center and joined the infantry. With a characteristic theatrical flourish, he telegraphed Hale: "Regret cannot accept your invitation. Am off to war."

Hubble was commissioned as a captain in the 343rd Infantry, 86th Division. He was subsequently promoted to major and was shipped off to France as a field and line officer. He arrived too late to see conflict (and it is therefore something of a mystery why in later discussions with colleagues he attributed a permanently stiffened elbow to a wartime shrapnel wound). In the summer of 1919, after a protracted stopover in England to catch up with friends from his "salad days" at Oxford, he returned to the United States and was mustered out of the army in San Francisco. He headed immediately to the Mount Wilson Observatory headquarters in Pasadena, still in uniform, to see Hale—to discuss with him the job at the observatory (twenty miles into the mountains behind Pasadena) originally promised in 1917.

Apart from service in the Second World War, Hubble would serve at the Mount Wilson Observatory until his death. Although he was to become California's best-known astronomer, he never tried to become an observatory director or sought a position of prominence in science politics. His commitment was to astronomical observation rather than organizational preeminence; even so he certainly seemed to enjoy the recognition that would later result from his research achievements, and he took to celebrity status as if born to it. Tales of Hubble have reinforced the "larger than life" aura he cultivated. He constantly smoked a pipe, and he favored a tightly belted army trench coat for external wear. When observing at a telescope he wore jodhpurs, high-topped military riding boots, and a Norfolk jacket (a belted single-breasted jacket with box pleats)—very much the attire of a "gentleman astronomer." Despite these carefully cultivated eccentricities, however, there is ample evidence of a degree of intolerance of colleagues, a demeaning attitude to those he considered of inferior intellect, snobbishness more characteristic of upper-class England than "middle America," and a tendency to be creative in his interpretation of the truth. In other words, Hubble was entirely human—with normal human frailties. Although he was blessed with an exceptional intellect and handsome features, any attempt to present him as superior in every way seems rather pointless. His contributions to astronomy speak for themselves, without the need to create for him (as many have tried to do) a posthumous persona of perfection at odds with reality.

During his time at Mount Wilson, Hubble turned the traditional vision of the universe upside down. Most of the astronomical discoveries of modern times pale into insignificance compared with what Hubble and his collaborators achieved there in the 1920s. This was a truly golden era of astronomy, when a new armory of giant telescopes, and increasingly sophisticated instruments, produced results that were both unexpected and profound in their implications for the nature of the cosmos. A cohort of dedicated scientists, looking at the cosmos in a new way, was merging "old" astronomy with "new" physics. The results would prove startling.

Remember that Harlow Shapley was still at Mount Wilson when Hubble returned from the First World War. Hubble and Shapley were never close. Shapley wrote that Hubble just could not get on with people. This certainly was not true—more accurately Hubble just could not get on terribly well with Shapley. It seems that Hubble resented the fact that Shapley had sat out the war in relative comfort at Mount Wilson, with plenty of access to time on the 60-inch telescope there, while Hubble had demonstrated his patriotism by volunteering for service. (For many years following his army discharge he would still introduce himself as Major Hubble.) But personal coolness between Hubble and Shapley was soon to turn to professional rivalry. Shapley was, of course, a principal proponent for the idea that the spiral nebulae lay within the Milky Way, which he argued was of such gigantic proportions as to represent the complete universe. But Hubble tended toward the notion that the spiral nebulae might be distinct island universes, and he was intent on finding the observational evidence to secure the hypotheses supported by Heber Curtis. The spiral nebulae were a source of fascination throughout his career. Thus Hubble and Shapley were on opposite sides of the Great Debate.

Hubble's investigation of the spiral nebulae passed through various phases. From 1922 to 1926 he pursued a classification scheme for nebulae (building in part on his doctoral work at the Yerkes Observatory but making full use of the power of the 60-inch and the recently completed 100-inch-diameter telescopes on Mount Wilson).

The majority of the nebulae Hubble classified were spirals, but a minority had a spherical or elliptical shape. (A small fraction had an irregular structure that belied a clear classification.) Hubble classified the nebulae according to a sequence that started with those of elliptical shape, then moved to the spiral structures with intertwined trailing arms of reducing tightness. But the spirals' sequence was split into two classes—those nebulae in which the spiral arms met at the center, and those in which the spiral arms linked to a short central bar (the so-called barred spirals). Hubble perfected his classification scheme throughout his career but wrongly sought an evolutionary path whereby a nebula starts life with an elliptical morphology, passing through a spherical phase before adopting one of two branches (one for normal spirals and the other for barred spirals). He then suggested that each branch passes through a tight spiral structure (the "a" type) that unwinds with time (to the "b" type and eventually the "c" type). It is now realized that Hubble's evolutionary ideas for nebulae were, quite simply, wrong. Nevertheless, his classification scheme was retained as a useful way of grouping nebulae with common characteristics.

One had to take care, in classifying elliptical nebulae, that the object was not just a spiral seen inclined at an angle to the line of sight. Edge-on spirals were easy to classify because of their disk shape with a central bulge, often with the distinctive dark strip down the middle signifying dust accumulated along the central plane of the object and blocking out light. And the face-on spirals were easy to identify from the intertwined spiral arms. The problem came with some nebulae lying between these two easy to identify extremes. Were they spirals seen at an oblique angle—or were they genuine ellipticals? Great care is always required in the classification of nebulae. And Edwin Hubble demonstrated that he was an observer and classifier with few equals.

Hubble had attended the conference in 1914 when Vesto Slipher announced his first results on the redshifts of spiral nebulae. The most obvious explanation for the reddening effect of the nebulae was thought to be the Doppler effect. Redshifted nebulae appeared to be moving away from Earth, with the extent of their redshift giving their velocity of recession. By February 1922 Slipher had been able to measure spectra for forty-one spiral nebulae—the vast majority displaying a redshift. Interestingly the Andromeda Nebula was blueshifted, implying that it was hurtling toward the Milky Way at a mind-boggling 300 kilometers per second! (It is now known that the Sun and other stars are rotating around the center of the Milky Way. Being swept along in this vast rotation, the solar system is at present moving toward the direction of the Andromeda Nebula at about 250 kilometers per second. Thus the velocity of approach of the Andromeda Nebula toward the Milky Way is actually a more sedate 50 kilometers per second.) Even with the new measurements of redshifts, there were still some astronomers who were intent on retaining the fiction that the spiral nebulae were gaseous clouds that had originated in the Milky Way. While they freely acknowledged that the redshifts implied velocities that were too great for the nebulae to now be part of the Milky Way, a range of increasingly fanciful proposals were produced to try to explain them away, for example the idea that nebulae were merely gaseous clouds from within the Milky Way that had drifted sedately to its periphery and had then been accelerated to great velocities by the pressure of starlight from the myriad of stars in the Milky Way. They also clung to Van Maanen's results. His claim that he had detected rotation of spiral nebulae—which would by impossible to detect at great distance—was so influential that many respected astronomers still forlornly hoped that the "island universe" hypothesis would not survive the test of time.

While working on the classification of nebulae, Hubble also searched for Cepheids in the spiral nebulae to determine their distances. In 1924, just three years after Shapley had left Mount Wilson to take over the Harvard College Observatory, Hubble secured the evidence he needed to prove that Heber Curtis had been right and Harlow Shapley had been wrong in the Great Debate on the nature of the spiral nebulae.

Hubble was able to use the 100-inch-diameter telescope on Mount Wilson, at the time the largest telescope in the world, to take a series of photographs of spiral nebulae. He found a Cepheid in the Andromeda spiral nebula in 1923. Within a year he had discovered twelve more Cepheids in the Andromeda Nebula, and using Henrietta Leavitt's period-luminosity law he was able to estimate that the Andromeda Nebula was at what then appeared to be a staggering distance of 2 million light-years, well beyond the outer bounds of even Shapley's expanded Milky Way. Other spiral nebulae were found to be at even greater distances. Here, at last, was the smoking gun—the long-sought-after evidence that the spiral nebulae lay at vast distances, well beyond the Milky Way. Kant's "island universes" were a reality. The impact on humans' understanding of the cosmos would prove to be genuinely profound.

Hubble seemed at first reluctant to publish his amazing results, although it is not obvious why, since he must have fully understood their implications. Did he harbor doubts about the Cepheid calibration? Was he waiting to get even more startling results? However, he did write to several people, including Henry Norris Russell and Harlow Shapley, describing what he had found. Shapley wrote back to Hubble saying that he had been amused by his report and merely acknowledged, at least initially, the discovery of variable stars "in the direction of Andromeda," perhaps in the forlorn hope that they might merely have been foreground objects within the Milky Way rather than embedded in the spiral nebula itself. But others realized that the evidence was irrefutable. Hubble was eventually persuaded by Russell that the results should be presented to a meeting of the American Association for the Advancement of Science to be held in Washington in December 1924. (The best paper would receive a prize of $1,000, so perhaps this was the inducement.) Hubble did not attend the meeting himself. Instead Russell read his paper to the meeting. There is a delicious irony here, since Russell had sided with Shapley in the Great Debate. Hubble's paper shared the prize for the best paper with one other researcher working in an entirely different field—just as Hubble had had to share his college scholarship!

The participants at the December 1924 meeting realized that the Great Debate was now over and a new era of enlightenment in cosmology had begun. Hubble's detection of Cepheids in several spiral nebulae had demonstrated that, without any doubt, they lay way beyond the outer bounds of the Milky Way and appeared to be independent galaxies in their own right. In truth this was not merely a new era of enlightenment in cosmology—this was a new era of enlightenment in understanding the sheer insignificance of the human condition in the grandiose structure of creation. The Hubble revolution matched the Copernican revolution in revealing the true nature of the cosmos.

Hubble's results were front-page news. This was the biggest story out of astronomy since the advent of mass communication, and the public was fascinated by what they read in the papers and magazines, saw on cinema newsreels, and heard on the wireless. The universe was suddenly dramatically larger than had been supposed, filled with "island universes" similar to our own Milky Way; and the new age of enlightenment seemed to be just perfect for the new era of social liberation and burgeoning prosperity. The general public could not learn enough about the excitement being generated in astronomy, and about the life of the handsome young astronomer from California. The new epoch of cosmic enlightenment was just right for the "Charleston generation," which wanted everything to be big, flashy, and extravagant—and Hubble had provided a big, flashy, and extravagantly populated universe. His timing was perfect.

The reaction from Heber Curtis on learning of Hubble's results was measured and calm rather than euphoric:

I have always held this view that the spirals are separate galaxies, and the recent results by Hubble on variables in spirals seems to make the theory doubly certain.

Shapley wrote to Hubble:

I do not know whether I am sorry or glad to see this break in the nebula problem. Probably both.

But Shapley was less forgiving in his comments to Harvard colleagues. Pointing to the letter from Hubble that informed him of the Andromeda Cepheid observations, he protested: "Here is the letter that destroyed my universe!"

Hubble was not entirely magnanimous in victory, and on more than one occasion made condescending comments about those who had suggested that the spiral nebulae lay within the Milky Way. Indeed he insisted on referring to external galaxies, the "island universes," as "spiral nebulae" for the rest of his life—long after the rest of astronomy started to use the generic term "galaxy" for an "island universe" (capitalized as Galaxy when referring to the Milky Way).

Actually Shapley could have solved the spiral nebula issue himself, years earlier, had he been more open to alternative explanation. In 1921 he had handed some photographic plates of the Andromeda Nebula to one of the Mount Wilson assistants, Milton Humason, to investigate for possible rotation of the nebula. Rather than finding any rotation (which would have demonstrated that the nebulae were nearby), Humason believed he had found some stars on the plates that appeared to vary in brightness and might just be Cepheids. He carefully indicated them with a marker pen on the reverse side of the glass photographic plates (that is, the reverse side to the precious photographic emulsion) and showed them to Shapley. They could not possibly be Cepheids, retorted Shapley, since as far as he was concerned the spiral nebulae were gaseous objects lying within the Milky Way. He turned the plates over, took his handkerchief, and carefully wiped the ink marks off them, thus also wiping away the potential for his own name to appear in the history books as the discoverer of the "island universes." Shapley later would find it hard to accept that van Maanen had made such a monumental error, reacting with disgust when the evidence of fault was finally produced: "I believed in van Maanen's results. . . . after all, he was my friend!" The moral of the story is that scientists must always seek independent validation of results—even if the initial results come from their best friend.

Scientific research always needs rigorous confirmation of all results ("search, research, and research again" is an effective mantra for science), even if personal friends present the initial results with conviction. Shapley had forgotten the discipline of confirmation needed in science and lost a unique opportunity to unravel the mystery of the spiral nebulae.

With his demonstration that the cosmos is truly vast, and made up of a multitude of independent galaxies lying well beyond the Milky Way, Hubble became an instant celebrity. He and his wife, Grace, were embraced by the social elite of California. If Hubble had achieved no more than this, then his place in the history books would have been assured. But the best was yet to come.

Having demonstrated that the spiral nebulae were separate galaxies, Hubble turned his attention to studying the properties of galaxies and their distribution within the cosmos. He wanted not only to determine the distances to the galaxies but also to understand the fact that their light was redder than it should have been.

Hubble, and Humason, who was now working as his assistant, set out to extend Slipher's earlier studies. Hubble promised Humason a new spectrograph for the task, and he honed his skills in the use of the new device. A careful and systematic approach typified the Hubble-Humason partnership.

Milton Lasell Humason is a fascinating individual in his own right. He was born at Dodge Center, Minnesota, on August 19, 1891. As a teenager he had gone on a summer camping holiday to Mount Wilson. He had so enjoyed the experience of mountain living that he persuaded his parents to let him drop out of school and return to the mountain to take a job as bellboy and handyman at the Mount Wilson Hotel. One of his jobs was to look after the pack animals. He was then given the job of driving the mule trains taking pieces of equipment up the mountain to build the new 60-inch-diameter telescope, a giant for its time. (In those days a mule train along a winding mountain path was the only mode of transport, a dramatic contrast to the highway and excellent mountain road one can now take to the observatory.) He met, fell in love with, and married the daughter of the observatory's chief engineer. The newlyweds went to live on a relative's ranch in La Verne. But Humason missed Mount Wilson, and his father-in-law managed to get him the job as observatory janitor and general assistant so that he could return. His interest in what the astronomers were doing blossomed, and before long his enthusiasm and dedication were rewarded: he was allowed to assist the astronomers with certain observing tasks. His attention to detail soon won praise, and before long George Ellery Hale made him an observing assistant; he particularly excelled at taking long photographic exposures of faint objects. His appointment as an observing assistant was somewhat controversial since he had no formal education and was the son-in-law of the chief engineer, but the initial fuss over his appointment soon died down as the Mount Wilson astronomers came to realize that they had a real talent in their midst. Shapley failed to benefit from that talent when he dismissed Humason's finding of Cepheids in the Andromeda galaxy. But Hubble and Humason worked well together. Hubble found the easygoing Minnesotan reliable and helpful, and not one to compete with Hubble's own status. Hubble was acknowledged as the senior partner in the enterprise. The two established an effective division of labor. Humason undertook most of the long observations, the endless hours at the telescope in the bitterly cold nights on the mountain. The uneducated bellboy, mule driver, and janitor would become one of the world's leading observational astronomers. Today's astronomers, armed with their multiple university degrees and advanced technology, would do well to reflect on the monumental breakthroughs made by their more humble forebear.

Very faint galaxies required extremely long exposures taken over many nights to gather sufficient light. The telescope dome was, of course, open to the cold night air. No form of heating was allowed, because this would cause convection, with warmer air rising around the telescope and causing star images to shimmer (an effect one readily observes with hot air rising off a tarmac). And of course no artificial lighting could be used, because it would compete with the precious light from the distant stars and galaxies. So Humason sat motionless, in the cold and dark, with his eye glued to the telescope's eyepiece, keeping crosshairs accurately positioned on a star used to guide the telescope. If the telescope tracking system wandered off the star by even the slightest amount as the Earth rotated, then Humason would have to speed up or slow down the telescope tracking to keep it pointing precisely in the same direction during the long exposure. Taking images of the night sky was challenging enough—but when the captured light was to be spread out into a photographic spectrum, then even longer exposures were needed to get sufficient light to produce something usable. Today professional astronomers jet between the world's leading observatories, where they sit in splendid comfort in control rooms viewing computer screens as their computer-controlled telescopes and advanced instruments search out the secrets of the cosmos. The contrast of this cushy existence of the modern observational astronomer with the hardships endured by earlier generations could not be more marked.

With Humason conducting most of the observations, Hubble worked in Pasadena at the observatory's headquarters, carefully analyzing the precious photographic plates that contained the faint images and spectra of the spiral nebulae. Humason, through patience and dedication, became the best observer of faint objects on Mount Wilson—a most remarkable accomplishment for a man with only a rudimentary education and no formal training in astronomy. Indeed, many of his contemporaries claimed he was the leading observational astronomer in the world in his heyday. During his career he took the spectra of an unprecedented 620 galaxies, most requiring many nights of precision observing.

While Hubble and Humason were conducting their experiments up on the mountaintop, theoretical physicists were conducting a lively debate on the nature and origin of the universe. The context was Albert Einstein's epoch-defining work on relativity. Einstein was the greatest scientist of the twentieth century—the individual who redefined the meaning of genius. Albert Einstein's impact on physics was as profound in his day as the great Isaac Newton's had been in his. The story of how a humble patent clerk, who had shown no great potential for mathematics in his youth, became the greatest theoretical physicist of the age is the stuff of fairy tales.

As a teenager Einstein had imagined what it would be like to ride a beam of light. In formulating his initial special theory of relativity in adulthood, he turned speculation into physical hypotheses. The theory forced scientists to think in an entirely new way, bringing new concepts into their descriptions of the physical world—such as time dilation (the slow running of clocks) and length contraction. Perhaps the most surprising outcome of his special theory of relativity was the equivalence of mass (m) and energy (e), such that mass can be envisaged as "frozen energy." The equation $E = mc^2$ (where c is the speed of light) is undoubtedly the most famous in science. The development of nuclear weapons later dramatically demonstrated the theory's validity. Einstein showed how the energy of a fast-moving object went into its mass, thus inhibiting it from traveling faster than the speed of light.

Before physicists had had time to absorb fully the implications of special relativity, Einstein challenged their intellects again in 1916 with a paper on the theory of general relativity, which linked the three dimensions of space with time, gravity, and matter. He had first presented his ideas to the Prussian Academy of Sciences in Berlin in November 1915.

General relativity envisaged space as a continuum that could be curved by matter, and matter's influence on this curvature of space was evidenced as gravity. An oft-quoted illustration involves a two-dimensional analogue. Imagine space as a sheet of rubber stretched out as a flat plane. If you roll a ball bearing across the stretched rubber, the bearing will travel in a straight line. But imagine now that you place a heavy metal sphere in the center of the rubber sheet, distorting the whole sheet. Now when you roll the ball bearing across the rubber sheet some distance from the center, the bearing will follow a curved path defined by the curvature of the rubber. This is clearly a simplistic explanation of the curvature of space by massive objects (planets, stars, and galaxies)—but it will suffice for our purposes here.

In general relativity Einstein anticipated that even light would be constrained to follow paths defined by the curvature of space. In 1919 a solar eclipse provided an early opportunity to test this hypothesis. The theory predicted that the Sun's mass would bend the light from stars viewed close to its rim, so that they would appear to be displaced in space compared with their observed position when they were well away from the visual direction of the Sun. Normally stars could not possibly be seen against the harsh glare of the Sun. But during the total solar eclipse of 1919 (when the Sun's light was fully obscured by the Moon passing in front of it), the distinguished British astronomer Arthur Eddington of Cambridge University measured the position of stars. Eddington confirmed that the starlight had been deflected by the Sun's gravity exactly as Einstein had predicted. General relativity moved instantly from the realm of an intriguing but unproved idea to the realm of scientific reality, and Einstein became a scientific icon. His disheveled appearance and wild hair aligned perfectly in the public mind with the image of the "nutty professor" from central casting.

It has been noted that general relativity requires "that space tells matter how to move—and matter tells space how to curve." The extreme state of space curvature is a black hole, the ultimate form of compaction of matter. Any astronomical body has what is called its "escape velocity"—the minimum velocity something must have to escape the body's gravitational field. The escape velocity for a black hole exceeds the velocity of light; that is, the gravitational field of a black hole is so intense that light itself cannot escape from it. (In the sheet of rubber analogy given earlier, the distortion of the rubber sheet is so great as to produce a hole in it through which any object passing nearby on the distorted sheet will plummet.)

Einstein was quick to realize that the general theory could be a powerful tool in cosmology for looking at the nature and origin of the universe, and he turned his equations to this task. But he struck an immediate problem. When he took the most reasonable-seeming description of the current universe and used his equations to predict its behavior, the solution he derived was unstable and implied that the universe would suddenly either expand or contract. Einstein's intuition was that the universe might well be expanding—but he realized that such a proposition would be unacceptable to his fellow scientists. (This was before Hubble's observations.) Although science had made enormous advances in understanding the nature of the heavens, conventional scientific wisdom demanded a static universe—a universe that had always existed. (But like Isaac Newton, the "father of gravity" some two hundred years earlier, Einstein was troubled by the fact that the action of gravity in a static but finite universe should be to cause all matter to collapse in on itself. Newton's conclusion was that the universe must be infinite, so that the net effect of gravity at any point was to keep the universe in balance. He left unanswered the conundrum that in an infinite universe any point will be subjected to the effect of gravity, however weak, from bodies all the way to infinity. Thus at any point in an infinite universe gravity would be infinite, which clearly is not the case. A static universe, either finite or infinite, produced some unanswerable questions.) Einstein decided that the only way to counter the apparently nonsensical implication of general relativity, that the universe was either contracting or expanding, was to introduce a special fudge factor into his equations. He called this term the cosmological constant, which could provide a hypothetical "antigravity" to counter any possible collapse. With the help of the cosmological constant Einstein could produce a solution for his equations that left the universe static. He should have had greater faith in his own equations. The "cosmological constant" would come back to haunt him. He later described it as the greatest mistake of his life—although intriguingly, in the most recent models of the expanding universe, the cosmological constant has made an unexpected comeback (more later).

Einstein's interpretation of his theory of general relativity was not the only show in town. The professor of astronomy at Holland's Leiden University, Willem de Sitter, had developed an alternative model of the universe based on his own interpretation of Einstein's equations. He found solutions to the equations in the absence of matter. Einstein was not impressed—a universe that would not allow the presence of a single star (or indeed even a grain of sand) hardly seemed to demand serious attention. As soon as any matter was introduced into de Sitter's model universe (even the tiniest amount), it started expanding uncontrollably. But de Sitter insisted that, so long as the amount of material in the cosmos was minute compared with the size of the cosmos, then his model must be seen as describing a very close approximation to reality. One of the implications of de Sitter's universe was that the properties of space itself changed with distance such that distant clocks would run slow and atoms emit radiation at redder wavelengths. Thus his model predicted that redshift would increase with distance, without demanding that distant objects be moving at extreme velocities. The de Sitter universe had an intrinsic mathematical beauty, even if astronomers struggled with its physical reality. And when news reached Europe from America of Slipher's redshift measurements for galaxies, then the de Sitter universe had to be taken seriously. It was in fact de Sitter's work that encouraged Eddington to undertake his 1919 eclipse expedition, which produced the experimental verification of the bending of light in a gravitational field predicted by general relativity.

There was another key player in the game of theories of the universe whose contribution, although profound, escaped the attention of most of the world of science. Aleksandr Friedmann was born in 1888 in St. Petersburg. His father was a ballet dancer and his mother was a pianist. His parents divorced and, unusually for the time, the father kept his son. The boy showed an early talent for mathematics, rather than displaying his parents' artistic traits. He studied at the University of St. Petersburg, getting involved in early discussions of quantum theory and relativity. In 1913 he was appointed to a position in the St. Petersburg Aerological Observatory, where he was to study meteorology. With the start of the First World War he asked to join an aviation detachment. He was soon flying aircraft on bombing raids but occasionally found time to pursue mathematics—trying, for example, to produce theories for the trajectories of bombs dropping. He wrote home to a colleague:

> My life is fairly even, except such accidents as a shrapnel explosion twenty feet away, the explosion of an Austrian bomb within half a foot, which turned out almost happily, and falling down on my face and head, which resulted in a ruptured lip and headaches. But one gets used to all this, of course, particularly seeing things all around which are a thousand times more awful.

Even his fighting gave him the chance to test his mathematical ideas. Again he wrote to a colleague:

> I have recently had a chance to verify my ideas during a flight over Przemysl; the bombs turned out to be falling almost the

way the theory predicts. To have conclusive proof of the theory I'm going to fly again in a few days.

Friedmann was decorated for bravery for his flying exploits. In 1916 he was appointed head of the Central Aeronautical Station in Kiev. With the Bolshevik Revolution in October 1917, the Central Aeronautical Station was closed, and Friedmann had to look for a new post. He lamented: "I'm very depressed; I often bitterly regret taking part in the war; it seems I achieved what I set out to do, but what's the use of it all now?"

Friedmann obtained a post of professor of mathematics at the University of Perm. But as the nation was thrown into civil war, further hardship followed. Perm was first occupied by the anti-communist White Army, which retained power until August 1919, when the Red Army took control and Friedmann fled the city. In early 1920, about the time Shapley and Curtis were engaged in the Washington Great Debate, Friedmann returned to St. Petersburg—the city's name had been changed to Petrograd—to take up a post at the Main Geophysical Observatory, and he also obtained a post teaching mathematics and mechanics at the university. Soon after returning to Petrograd he started taking an interest in the cosmological implications of Einstein's theory of general relativity. It was the mathematics rather than the astronomy that excited him.

The Friedmann cosmology unashamedly promoted an expanding universe. No excuses—and no fudge factors intriguingly named to disguise ignorance. Friedmann had produced a clear prediction demanding observational confirmation or rejection. Einstein was unimpressed and challenged Friedmann's calculations—wrongly, as he later graciously admitted. But Friedmann's paper explaining his cosmology was largely overlooked and was not rediscovered until very much later, when the evidence for an expanding universe had finally been obtained. One reason why Friedmann's paper was overlooked was his untimely death. In July 1925 he undertook a record-breaking high-altitude balloon flight to 7,400 meters to conduct meteorological and medical experiments. The experience weakened him severely, and on his return to Petrograd—which by now had been subjected to yet another name change, to Leningrad—he was unable to cope with a subsequent bout of typhoid. Had he lived and been able to promote his cosmological theory, then perhaps it would not have escaped attention for so long.

The curvature of space implied by general relativity allows for a situation that appears at first to be far from intuitive—namely that the universe could be finite (that is, there is only so much of it) but unbounded (that is, it has no outer limit) and is without a center. Imagine an ant constrained to live on a piece of paper, its two-dimensional "universe." This universe is finite (the paper has a certain size), but it clearly has a boundary (the edges of the paper), and it also has a center (the center of the paper). Even if the piece of paper is replaced with a piece of rubber that can be stretched to give the ant an "expanding universe," then it is still finite, bounded, and has a center. But general relativity allows space to curve. So imagine that we take the piece of paper representing the ant's two-dimensional universe and wrap it around a nice round grapefruit (in essence we have added an extra dimension to the illustrative two-dimensional universe). Now the edges of the paper overlap, so the ant can wander over its "universe" without encountering a boundary and without knowing (or needing to know) where the center of the paper universe is. Space curvature has given the ant a finite, unbounded universe without a center. And the intriguing thing about such a universe is that if you head off in any direction and keep going straight, you end up back where you started. We could make such a universe expand (imagine the ant living on the surface of a balloon that is being blown up). In this simple illustrative example of the ant's universe we have taken a finite, bounded universe with a center and, by giving the space curvature have converted it to a finite, unbounded universe without a center (which can even be allowed to expand). General relativity and space curvature can produce a universe with all sorts of interesting and counterintuitive properties.

In the summer of 1928 Edwin Hubble visited Leiden to chair an international meeting on nebulae. Following his return from Leiden, and perhaps inspired by discussions with de Sitter, Hubble started his campaign with Humason to look at the relationship between the redshift of galaxies and their distance.

By 1929 Hubble had studied twenty-four galaxies, some with spectra obtained by Slipher and others being new results from Humason. All the twenty-four spiral galaxies observed showed redshifts of varying degree. The majority of these galaxies were too distant and faint to allow a search for Cepheids. So instead, for those without Cepheids, Hubble concentrated on what he believed were their brightest stars. He assumed that the brightest stars in different galaxies might all have the same intrinsic brightness (a not unreasonable assumption), so that he could then estimate the distance of the galaxies through the inverse square law once a cross-calibration could be secured from galaxies whose distance had been determined from the use of Cepheids. The most distant galaxy of the twenty-four studied was in the Virgo cluster of galaxies, where Hubble had measured a large redshift and used the "brightest star" to estimate a distance of 6.5 million light-years.

Hubble's use of brightest stars as a distance yardstick, although intuitively acceptable, did rather lead him astray in his investigations. The problem was that even with the mighty new 100-inch diameter telescope, it was not always possible to resolve single stars in the distant galaxies. Hence what might look like a bright single point implying a brilliant star could in fact be a close cluster of less bright stars, giving the appearance of a single bright star in an unresolved image. The other possibility was that the bright point was in fact a glowing cloud of gas rather than a star. It would later turn out that such ambiguities produced some erroneous distance estimates in Hubble's work at this time. Later, with more powerful telescopes and the application of the stellar classification techniques pioneered by Annie Jump Cannon, some of these ambiguities would be resolved.

The twenty-four galaxies for which Hubble had redshifts and estimated distances did, nevertheless, reveal a spectacular result, which he reported in a paper published in 1929. There appeared to be a linear relationship between a galaxy's radial velocity, derived from its redshift, and its distance, estimated from identifying

Cepheids or from the brightest stars. Thus a galaxy twice as far away as another appeared to be traveling with twice the velocity. The speed of recession was related to distance by a constant that now carries Hubble's name—the "Hubble constant." Because it seems likely that the Hubble "constant" has changed over the history of the universe, some prefer to call it the "Hubble parameter."

An extra yardstick had now been provided to estimate the distances to the most distant of galaxies. Parallax, Cepheids, redshifts, and other verifying methods enabled cosmic distances to be estimated with sufficient precision to allow studies of the structure and evolution of the universe to be pursued with enthusiasm and increased certainty.

Hubble's result, that the speed of recession of galaxies is proportional to their distance, is often presented as an unexpected piece of serendipity. That simply is not the case. Such a relationship had been proposed previously, not only from the de Sitter model universe, but also from one of Slipher's observations that galaxies that appeared smaller on the sky (and were presumably therefore at greater distance) showed larger redshifts. Thus Hubble and Humason were looking deliberately for just the relationship that some later commentators presented as being merely an accidental discovery. Indeed Hubble's widow later confirmed that Hubble's search for a redshift versus distance relationship was inspired by his discussions with de Sitter in Leiden.

Eager to confirm their results, Hubble and Humason pressed on to even greater distances, taking longer and longer exposures of fainter and fainter galaxies. By 1931 they had reached the Leo cluster, which Hubble estimated was 105 million light-years away. (It was traveling at what at the time appeared to be a quite staggering velocity of 19,700 kilometers per hour.) Many astronomers had difficulty accepting such astonishing velocities and distances.

It has often been pointed out that Hubble's results of 1929 did not really demonstrate a clear relationship between redshift and distance, because of the large scatter in the results. It has been implied that perhaps Hubble, swayed by his discussions with de Sitter, forced the sort of relationship that he wanted to see from meager data. Such a claim is probably unfair. It is more likely that

Hubble already had preliminary results for more distant galaxies (which he would eventually publish in 1931) that more clearly indicated the redshift-distance relationship in the form he published it.

There were some who argued that the distances implied by Hubble from his redshifts, although impressively large, must actually represent significant underestimates. Their argument went something like this. Taking the angular extent of a typical spiral galaxy Hubble was measuring, and using his calculated distances, then an estimate could be made of the diameter of the galaxy in light-years. Results of a few tens of thousand of light-years were found. Such diameters were startling enough, without doubt. But they were considerably less than the diameter that had been calculated for the Milky Way. Now there was no real reason to suppose that the Milky Way should in any way be special—such as being bigger than all other measured galaxies. Would it not be more reasonable to assume that the Milky Way was merely "average" on the cosmic scale of things—in the same way that after Copernicus we had to accept that there was nothing special about planet Earth, and after Shapley we had to accept that there was nothing special about the solar system? If the Milky Way was merely "average," then this would imply that the distances being inferred by Hubble, however enormous they might appear to be, were short by a factor of two or three. Of course this was a somewhat obscure and indirect argument, but it could not merely be ignored. This type of argument is based on the so-called principle of terrestrial mediocrity. Astronomers are wise to work from the starting point that there is nothing special about humans, the Earth, the Sun, the Milky Way, or our Local Group of galaxies. In the grand order of things, we can best be thought of as no better than mediocre. Hasn't the history of the evolution of human thinking demonstrated just this fact?

With the galaxies traveling at such amazing velocities as those implied by the Hubble and Humason results, could it be that the universe as a whole was expanding? The implications of that possibility were at once both staggering—and frightening.

The fact that all galaxies are receding from the Milky Way does not mean that the Milky Way is at the center of the universe but rather that the Milky Way, and all other galaxies, can be viewed as part of a general universal expansion. In such a universal expansion, it would not matter which galaxy one was part of; all other galaxies would be viewed as expanding away with a velocity dependent on distance (in accordance with the Hubble relationship). This is sometimes called the "cosmological principle." Imagine guests crowded uniformly together at a cocktail party. To relieve the pressure the host opens several adjacent rooms so that the guests can drift apart, uniformly distributing themselves with a more comfortable spacing. Your most immediate neighbor may have moved away from you by only a few feet, whereas more remote guests are now fully a room away. Thus one has a version of the "cosmological principle" for cocktail parties. As the guests distribute themselves more comfortably, they all see and experience a similar general expansion whereby guests nearby appear to have moved away just a small distance and guests farther away have moved on by a greater distance.

There is a point that needs to be emphasized about the redshifts of galaxies. What the redshifts are implying are *apparent* velocities of the galaxies, not the real motion of galaxies through a preexisting space of dimensions independent of the presence of the galaxies. These cosmological redshifts are caused by the stretching of space itself, rather than by relative motion within a preexisting space (as the conventional interpretation of the Doppler effect would imply). Put another way, in the time the light from a distant galaxy takes to reach Earth, the space between the galaxy and Earth has increased—which leads to the wavelength being stretched by a commensurate amount. Nevertheless, true Doppler shifts are useful for measuring the relative motion of astronomical objects, for example the motion of galaxies within a cluster with respect to their common center of mass, or the motion of stars in orbit around each other or around the center of a galaxy. But the cosmological redshift needs to be considered as a distinct phenomenon resulting from the expansion of space itself.

Einstein was overjoyed when he heard of Hubble's results. His cosmological constant, the fudge factor he had introduced to produce a static universe from general relativity, appeared to be entirely redundant after all. His only mistake seemed to be to have doubted his initial intuition that his theory demanded an expanding universe. In 1931 he made a special pilgrimage to the top of Mount Wilson to thank Hubble personally. (Seventy years later Einstein's pilgrimage to the mountaintop would be revealed as having been unnecessary, as the cosmological constant reemerged into mainstream research in an unexpected way.)

Determining the value of the Hubble constant with some certainty proved to be an enormous challenge for science, and there has been a continuing dispute as to its correct value. From the beginning estimates varied by a large factor, and a reduction in this uncertainty awaited observations from space almost five decades after Hubble's death, as we will later recount.

Edwin Hubble was one of the first "superstars" of modern science. In his day probably only Albert Einstein had a comparable profile with the public. Hubble was a celebrity on many fronts. During the 1930s and 1940s he was the toast of Hollywood, on the "A-list" for society parties. He was a friend of Charlie Chaplin, William Randolph Hearst, and Helen Hayes, and a confidant of Aldous Huxley. In February 1948 his portrait graced the cover of *Time* magazine. Various dignitaries and celebrities braved the perilous journey up Mount Wilson to see the famous astronomer at work. Hubble enjoyed his celebrity status and courted publicity. Fame sat comfortably on his lanky frame.

Hubble was also a brilliant writer. The impact of his published works depended as much on his mastery of language, and his elegant expository style, as on the actual scientific content of his papers (which was, of course, substantial). Few scientists could match his power of presentation or his clarity of expression.

In the summer of 1942 Hubble again answered the call of his country to join the war effort, but this time far from the field of combat. He was appointed head of the Aberdeen Proving Ground in Maryland, testing new ordinance (a task far from risk-free). For his war work he was awarded the Medal for Merit in 1946.

The only major honor to escape Edwin Hubble was a Nobel Prize—and it was not for want of trying on his part. It is rumored that he hired a publicist to promote his cause with the Nobel committee, although the committee was still struggling with the issue of whether prizes for astronomy could fall within the category for physics. Insiders say that he was on the verge of gaining this ultimate accolade when he died, at sixty-three, from a cerebral thrombosis on September 28, 1953. He had had an earlier heart attack but had eased back into his observing schedule and other commitments and had seemed to make a full recovery. His death was unexpected. Having complained to colleagues of feeling unwell, he drove home—and dropped dead on his driveway.

Hubble's wife, Grace, organized a private cremation ceremony, and the great man's ashes were scattered at an undisclosed location. For his many friends and admirers there was no funeral service— no public ceremony. Hubble was there one day and gone the next. Astronomy was at a loss for how to mourn. The quiet Minnesotan Milton Humason, who had obtained so many of Hubble's observations, died suddenly at his home near Mendocino, California, on June 18, 1972.

Hubble would have taken great satisfaction from one honor that was bestowed on him posthumously. The Space Telescope was launched April 24, 1990, on the Space Shuttle *Discovery*. It is the most expensive and sophisticated astronomical experiment ever built, and once in orbit it was named the Hubble Space Telescope in honor of the greatest observational astronomer of modern times. And the Hubble Space Telescope would eventually help determine a definitive value for Hubble's "constant."

Chapter 5

THE NATURE OF CREATION

I t is tempting to extrapolate the expansion of the universe, indicated by the redshifts of the galaxies, back to a time billions of years ago when the galaxies would presumably have been tightly packed together. This thought experiment is rather like running a video backward—trying to imagine what the universe might have looked like a billion years ago, five billion years ago, or even ten billion years ago. The Belgian cleric Georges Lemaître was one of the original proponents (in the 1930s) of the idea that the matter of the universe was originally concentrated in a superhot, superdense form. Then he imagined a glorious epoch of creation that initiated the rapid expansion of this dense primordial system.

Born in 1894 in Charleroi, Belgium, Georges Lemaître was just nine years old when he precociously indicated to his startled, but nevertheless suitably impressed family that he wanted to become both a priest and an astronomer. The reason, he later noted, was that "there are two ways at arriving at the truth. I decided to follow them both." Lemaître was studying civil engineering at the Catholic University in Louvain when the Germans invaded Belgium at the start of the First World War. He immediately volunteered for army service and saw the worst of conflict, being involved in hand-to-hand fighting. He was awarded the Croix de Guerre *avec palmes* for his bravery. Following the armistice he restarted his university studies, in 1923 joining a seminary to become a priest. He visited Cambridge, England, to spend a year studying with Arthur Eddington, whose prominence had recently been enhanced by his observational proof of Einstein's theory of general relativity. Harvard was Lemaître's next port of call, where he worked with Harlow Shapley. While in the United States he visited both Slipher and Hubble, so was well versed in the controversy of the spiral nebulae and the early measurements of their redshifts. Interestingly he was at the meeting in Washington where Russell announced Hubble's results for the distances to the spiral nebulae, effectively bringing a satisfactory conclusion to the Great Debate of 1920.

Lemaître returned to Belgium in 1927 to become professor of astronomy at Louvain. He was enthused by the implications of Hubble's work and much else he had heard while in the United States. The year of his return to Belgium he published a paper that would become a classic in astronomy, in which he used Einstein's general relativity equations to propose ideas for the origin and expansion of the universe. However, it was published in a little-known Belgium journal and largely escaped the attention of the mainstream of scientific thought. Einstein was initially dismissive of it; he patronizingly observed that Lemaître's mathematics was acceptable but that his physics was appalling. And the paper may have languished in obscurity had it not been for a chance remark at a meeting in London.

During the meeting in January 1930 of the Royal Astronomical Society (the bastion of astronomy in Britain) the topic for debate was the implications of the theory of general relativity for cosmology. The Einstein and de Sitter models remained the focus for current thinking—Friedmann's paper, like Lemaître's, having escaped attention. Both Einstein and de Sitter had forced their cosmologies to a static universe (a universe that was neither expanding or contracting) because that was what current scientific conventions demanded. Arthur Eddington chaired the meeting. Participants were left in the usual state of confusion over the Einstein model, which required an unexplainable fudge factor (the cosmological constant) and could not deal with the redshifts. The de Sitter model, on the other hand, accommodated redshifts but worked only for a universe strangely devoid of any matter! The situation was most unsatisfactory. The theoreticians may have been playing with eloquent mathematics, but there seemed to be no physical reality on which to base their speculations. Eddington noted: "One puzzling question is why there should be only two solutions. I suppose the trouble is that people look for static solutions."

Eddington's comment found its way to Georges Lemaître in Belgium, and he wrote to his old tutor referring him to his own solution of Einstein's equations, which described an expanding universe that had had its origin in a single moment of creation. Although Eddington doubted the creation aspect of Lemaître's hypothesis, he was excited by the implications of his paper and its mathematical clarity, and wrote to the journal *Nature* publicizing its importance. Suddenly Lemaître's forgotten scholarship was front page news. And at about this time Friedmann's earlier, but also forgotten, expanding universe solution of Einstein's equations was rediscovered and given serious consideration by scientists. Friedmann had, of course, been a mathematician, and his work reflected his entrancement with the beauty of mathematics rather than his understanding of any supporting astronomy. (His paper made no reference to Hubble's work on the spiral nebulae or Slipher's redshift observations; indeed it is not obvious he even knew of them when he undertook his research.) By contrast Lemaître was an astronomer, and he was able to fully understand the cosmological implications of the inference that space could stretch and carry the galaxies along with the expansion. He also knew what observational evidence would be available to check this hypothesis.

But if the universe was expanding, what could it be expanding from? Friedmann's model suggested that the universe had originated in a point—"a singularity." But Lemaître postulated the existence of a single supermassive "primordial atom," comprising all the material and energy from which the universe would eventually be made up.

It was known that large naturally occurring atoms such as uranium were unstable and broke up into more stable light atoms by the process of radioactivity—the emission of energetic particles from the nuclei of atoms. Lemaître pictured an extremely complex variant of radioactivity as the mechanism by which the constituents of the universe were formed from the primordial atom. The details of this "super-radioactivity" could only be speculated about, although he proposed that the first stage of decay would be from the dense primordial atom to star-size objects. From the standpoint of our current knowledge of physics, the notion appears to be absurd. But Lemaître was working within the limitations of the science available to him, and in the absence of anything else to explain a creative event he had to resort to his imagination. Lemaître thought that the universe probably did go through an initial expansion, followed by a static phase, before the present epoch of expansion was triggered. He explained his idea thus: "We could conceive the beginning of the universe in the form of a unique atom, the atomic weight of which is the total mass of the universe. This highly unstable atom would divide into smaller and smaller atoms by a kind of super-radioactive process."

In December 1932 Lemaître visited Pasadena to discuss his ideas with Hubble and Einstein. He gave a well-attended lecture where he explained his cosmological model. This time Einstein was somewhat more polite.

The Lemaître universe not only generated interest in the highest echelons of science; it was also greeted with intense interest in the highest echelons of the church, especially since the new theory for the origin of the universe required an epoch of creation that accorded so well with the philosophy of Genesis. And what's more, the theory had been produced by a scientist who could be thought of as "one of their own." Lemaître was made a member of the Pontifical Academy of Sciences in 1936, and he would later serve as its president. It seemed that centuries of mistrust between science and the church might finally be put to rest, following the works of Copernicus, Galileo, Darwin, and many other eminent scientists who had, usually inadvertently, challenged Christian dogma in various ways (albeit motivated by the pursuit of scientific truth, and certainly not malicious intent against the church).

Although Lemaître's expanding universe model, based on solving Einstein's equations, found favor with fellow scientists, his notion of a primordial atom certainly did not. It conjured up the notion of the supermassive "atom" decaying into a space that already existed. This, of course, was not what Lemaître had intended. But the primordial atom, although popularized with the general public, was never given serious consideration by the physics establishment. It just did not fit with the known theories of physics. There had to be some other way of dealing with the expansion of the universe.

For some unknown reason Hubble had not quoted Lemaître's work in his 1929 paper describing the relationship between redshift and distance, despite Hubble's knowing of Lemaître's work from the latter's earlier visit to see him. Hubble did refer to de Sitter's work, however, presumably because of their Leiden meeting.

In his 1929 paper Hubble quoted a value of 525 kilometers per second per megaparsec for the constant relating redshift and distance, although the imprecise quality of the data meant that many people felt that the precision of the estimate was to be seriously questioned. (The megaparsec, which equals a million parsecs, or 3.26 million light-years, is the preferred unit for the extreme distances of galaxies. Since the Hubble constant is velocity divided by distance, the unit used for it is kilometers per second per megaparsec.) However, in their 1931 paper called "The Velocity-Distance Relationship among Extra-Galactic Nebulae" (note the use of "Extra-Galactic Nebulae" rather than "Galaxies"), Hubble and Humason had added redshifts for a further fifty galaxies to the earlier list, thus more than doubling the number from the 1929 paper. Their largest redshift was now 20,000 kilometers per second, placing the galaxy in question at an estimated distance of 100 million light-years. The universe just kept on growing! Where could it possibly end? Within a mere decade humanity had had to come to terms with the realization that, not only were there other "island universes" beyond the Milky Way, but they were part of an expanding universe of truly colossal extent.

Einstein and de Sitter joined forces in 1932 to produce a new cosmology based on the theory of general relativity. This time both accepted an expanding universe. The Einstein–de Sitter model set the curvature of space to zero. It predicted that the current expansion of the universe would gradually slow over the eons—without ever quite coming to a halt and then eventually collapsing under the influence of gravity. This state would require a density of material in the universe such that the strength of gravity trying to halt its expansion was exactly balanced by the energy of motion of the expansion. However, to achieve this density would require a larger amount of matter than could be seen in all the visible stars, star systems, and galaxies—that is, matter that did not emit light or other forms of detectable radiation. This so-called "dark matter" was later implied to exist from observing its gravitational effects on galaxies and the light from galaxies, and the search for dark matter has been a continuing source of fascination for astronomy, as we shall see in the next chapter.

The 1929 paper by Hubble had postulated a value for the "Hubble constant" of approximately 500 kilometers per second per megaparsec. If it is assumed that the universe has always expanded at the current rate, then by mapping the expansion of the universe back through time at this rate one gets to an initial point in time when all the material in the universe would have been packed tightly together. This "time" is the reciprocal of the Hubble constant (that is, it is calculated by dividing one by the Hubble constant). Of course it is a big assumption to say that the universe has always expanded at its current rate. Perhaps the expansion has been slowing gradually, as the Einstein–de Sitter model suggests. If it was indeed expanding more rapidly in the distant past, then it would have reached its current size more quickly that the present rate of expansion would imply. That is, using the present rate of expansion could slightly overestimate the true age of the universe. Or perhaps, even, the expansion has been accelerating gradually, in which case the true age of the universe would be somewhat greater than that inferred from the present rate of expansion, since in the past the expansion would have made slower progress than it is making at present. Because of this uncertainty it would be wrong to assume that the Hubble "constant" gives the true age of the universe. Instead we refer to the "Hubble time" as being merely indicative of the age of the universe. A Hubble constant of 500 kilometers per second per megaparsec implies a Hubble time of about two billion years.

Although the implied Hubble time was an impressively long period, it was unacceptable to science as indicating anything like the true age of the universe (especially if the expansion was slowing down, as many cosmologists then supposed). The reason was that another branch of science was already doing a fine job in estimating the age of planet Earth.

By looking at the rate of decay of radioactive matter in rocks, geologists were making great strides in estimating the Earth's age. Their method was based on the assumption that much of Earth's lead had been formed by the radioactive decay of uranium within the Earth's crust. To estimate the amount of lead within the Earth at the time of its formation, it was necessary to look at meteorites—scraps of debris from the formation of the solar system that survive passage through the atmosphere and can be used for analysis of the composition of the nascent solar system. By careful calculation of the amount of lead that the early Earth must have contained, based on the analysis of the composition of meteorites, the extra amount that must have been added by the radioactive decay of uranium could be inferred. Then the time for such a buildup of additional lead could be calculated knowing the rate at which uranium undergoes radioactive decay. And this figure was in excess of four billion years. There could be no doubt. The radioactivity data could not be fudged—the Earth had to be at least four billion years old. The evidence was irrefutable. So how could the cosmology derived from Hubble's observations suggest that the universe was only two billion years old? The cosmology must be nonsense; there seemed to be no other possibility. Hubble's reputation was suddenly on the line.

Stellar astronomy was also producing problems for the cosmologists. Astronomers studying globular clusters of stars in the Milky Way had reasoned that they could estimate the age of the Milky Way by a careful analysis of the distribution of the brightness and colors of the thousands of stars in a cluster. Globular clusters are thought to contain the first generation of stars formed when a galaxy is born. All the stars in any cluster would have been formed at roughly the same time. The more massive stars in the cluster would be the first to use up their hydrogen and helium and become what is called red giants (large, bloated stars nearing the end of their life), while less massive stars would still continue to shine normally. By looking at the relative populations of red giants compared with stars of lesser mass, it is possible to infer an approximate age for a globular cluster (and therefore for the Milky Way). But the astronomers undertaking these studies were getting a mind-boggling figure of eighteen billion years for the age of clusters. A later recalibration could bring this figure down only toward about twelve billion years—but no further. Reassuringly the age inferred for the Milky Way from stars in globular clusters was comfortably greater than the four-billion-year age of the Earth and therefore the Sun, as of course was required, since the Sun is not thought to be a first-generation star within the Milky Way but must have been formed sometime after the Galaxy came into being. But by comparison with the inferred globular cluster ages, a low Hubble time of two billion years looked (at best) bizarre.

With the radioactivity data and the globular cluster data at least giving a consistent picture, the finger of suspicion pointed firmly in the direction of the Hubble constant measurements. So what could explain the discrepancy between the results obtained from geology and stellar astronomy versus those from cosmology based on redshifts? Perhaps the results of Hubble and Humason were wrong. But the two astronomers' reputation for careful observation and interpretation was unimpeachable. Lemaître had suggested an indefinite period when the universe was static before it started expanding, and maybe there was something in that hypothesis that could offer salvation for the expanding universe idea and the age implied from redshifts. Or perhaps redshifts did not imply expansion, and an entirely new theory of physics was called for. Some astronomers argued for the existence of "tired photons"—photons that moved to red wavelengths as they traveled over the eons through the vast distances of intergalactic space. Whatever the explanation might be, the excitement of Hubble's results, and Einstein's and Lemaître's theories, was now being seeded with considerable doubt.

It would be twenty years before significant progress was made to resolve the anomaly, and the solution would again require a mix of old-world ingenuity and new-world technology. Walter Baade was a German, born in 1893, who had emigrated to the United States in 1931 to work at the Mount Wilson Observatory. He had not got around to taking out U.S. citizenship until 1940, and formalities had not been completed when the United States entered the Second World War. As a German citizen he was initially subjected to an 8:00 P.M. to 6:00 A.M. curfew, thus stopping him from using the telescopes. But eventually the authorities relented, realizing that his allegiance was now entirely with the United States, and he was able to return to his observing routine on Mount Wilson. With Hubble and other astronomers away on war service, he had plenty of access to the 100-inch telescope (still at this time the world's finest). What is more, the blackout in the cities in the valleys below produced ideal conditions for astronomy. Baade was keen to follow up Hubble's work on the Andromeda galaxy, and he took numerous long-exposure photographic plates and spectroscopic data of stars and nebulae within that spiral galaxy. This research, aided by Annie Jump Cannon's classification scheme, revealed that the stars in spiral galaxies (including our own, the Milky Way) fell into two main populations. Those stars lying away from the center, in the spiral arms, were younger and contained a rich variety of elements; they were labeled "Population I" objects. Those stars in the central nucleus of the galaxy, and also those lying in the globular clusters forming a spherical halo around the galaxy, were older and made up almost entirely of hydrogen and helium; these stars were called "Population II" objects. We now know that the Population II stars are the first-generation stars in a galaxy and the Population I stars have been formed in the spiral arms more recently from material that has already undergone a degree of nuclear processing by earlier-generation stars.

Cepheid variables were found to be different in the two populations. Population I Cepheids (now known as classical Cepheids) were much the brighter. Population II Cepheids (now called W Virginis variables) were much fainter than classical Cepheids.

In 1948 Walter Baade was able to transfer his research on the Andromeda galaxy to the new giant 200-inch-diameter telescope recently opened on Palomar Mountain. This telescope was George Ellery Hale's final and finest funding triumph. Hale had written an article for *Harper's Magazine* titled "The Possibilities of Large

Telescopes." The opening sentence read: "Like buried treasures the outposts of the universe have beckoned to the adventurous since immemorial times." Rumor has it that the multimillionaire industrialist and philanthropist J. D. Rockefeller read only the first few sentences of the article before ringing Hale with an offer of $6 million from the Rockefeller Foundation to sponsor a new telescope. This inspired patronage would eventually bring forth the mighty 200-inch telescope, following a delay caused by engineering difficulties and the Second World War.

On the 200-inch telescope Baade was particularly interested in searching for a class of variable star known as RR Lyrae stars. RR Lyrae stars have a similar behavior to Cepheids but are much fainter and their periods are shorter, ranging from a few hours to a couple of days. Studies in the Milky Way had shown that these variable stars were particularly useful in providing accurate distance estimates, and RR Lyrae stars were assumed to exist in other galaxies as well. Although Hubble had not been able to detect RR Lyrae variable stars in even the nearest of the spiral galaxies using the 100-inch telescope, Baade was satisfied that even at the distance of one million light-years that Hubble had calculated for Andromeda, RR Lyrae variables would be detectable with the increased power of the 200-inch telescope. However, they were nowhere to be found. Repeated searching with longer exposures down to fainter and fainter limits did not reveal the expected RR Lyrae stars. The only possible explanation was that the whole Cepheid calibration scale must be wrong and Andromeda must be much farther away than Hubble had estimated. And if Andromeda was farther away, then a key point on the redshift-distance curve was wrong; and hence the Hubble constant and the inferred age for the universe must also be wrong. This looked like a case of going back to the drawing board!

The error was eventually tracked back to Shapley's analysis of Cepheids, done some thirty years before Baade's Palomar observations. Not that the error resulted from carelessness, since Shapley had been a most careful observer of real ability. But Shapley did not have the knowledge that astronomers now had about the two distinct populations of stars. It turned out that dust in the plane of the

Milky Way, and the fact that there are two distinct populations of Cepheids, colluded to fool Shapley. The Population I Cepheids, although intrinsically brighter, lie in the plane of the Milky Way, where dust dims their brightness. By contrast the fainter Population II Cepheids, the so-called W Virginis stars, lie in globular clusters away from the plane of the Milky Way and are therefore not dimmed by dust. Shapley had been keen to gather all the Cepheid data he could and, not being aware of two distinct populations, had naturally assigned them all to a common grouping. The dust's effect by chance made the classical Population I Cepheids appear similar to the fainter Population II Cepheids, thus leaving Shapley entirely unaware that they were of fundamentally different intrinsic brightness.

Hubble had used the brighter classical Cepheids in his work estimating the distance to the Andromeda Nebula. That was not the problem. What was the issue was that the classical Cepheids had been incorrectly classified along with the fainter Population II Cepheids in setting the period-luminosity relationship used to estimate distance. Thus the stars Hubble had used in Andromeda were in fact intrinsically much brighter than Hubble had supposed, and Andromeda was as a consequence at twice the distance Hubble had estimated—that is, it was at a distance of 2 million light-years. And if Hubble's key reference point moved by a factor of two, then the Hubble constant could at a stroke be halved to 250 kilometers per second per megaparsec. And so the resulting estimate of the "age" of the universe, the Hubble time, increased to four billion years—still not nearly long enough but more in line with the estimate radioactivity gave for the age of the Earth. Things suddenly looked more promising, although the twelve-billion-year age estimated for globular clusters still represented a source of embarrassment. But the story was far from over, since the Hubble constant would continue to shrink in the decades to come.

Physicists meanwhile continued to resist Lemaître's vision of a primordial atom. The whole idea of a creative event was anathema to many scientists. Despite having personally promoted Lemaître's work, Arthur Eddington remarked: "The notion of a beginning is repugnant to me." In the atheist Soviet Union, under the tyrannical rule of Joseph Stalin, any suggestion of a creative event was also repugnant. Freethinkers in cosmology in the Soviet Union were caught in successive purges, essentially ensuring that a generation of Soviet scientists was lost to cosmology—at least in their homeland.

If the universe age deadlock was to be broken it would require someone of immense imagination. George Gamow was born in Odessa in the Ukraine in 1904. He lived through the turmoil of the Russian Revolution and civil war, studying physics at the University of Leningrad. In 1931 he was appointed master of research at the Academy of Sciences in Leningrad, at just twenty-seven years old. Gaining such a prestigious post at such a young age was recognition of his brilliance. Gamow found the repression of the Soviet Union intolerable, and he and his wife tried to escape many times (but by good fortune their escape attempts were not detected). When he was given permission to attend a scientific conference in Brussels in 1933, he sought an audience with Vyacheslav Molotov, Stalin's right-hand man, and persuaded him that he should be allowed to take his wife along to the conference as his "personal assistant." Permission was granted. Not surprisingly, after the Brussels conference Gamow and his wife did not return to Russia but traveled on to the United States, where he secured an appointment at the George Washington University in Washington, D.C. With his vivid scientific imagination and mischievous sense of fun he quickly became a popular member of the U.S. science scene. In the 1940s he became well known through his *Mr. Tompkins* books explaining the complexities of science (including relativity) to the general public. For the fun-loving Gamow science was a joy, and he just had to share the excitement with others.

Gamow looked for an alternative to Lemaître's "primordial atom" theory. As a physicist rather than an astronomer, he came to the creation problem from the opposite direction. Instead of imagining backward in time from the current expanding universe to Lemaître's concept of a massive primordial atom, Gamow thought from the beginning forward (or from the bottom up). That is, he thought about building the very first atoms from subatomic particles. Although atoms are the fundamental building blocks of all matter, we know that they comprise a dense central nucleus containing subatomic particles called neutrons and protons, around which minute particles called electrons orbit. Gamow's universe started with a hot, dense "soup" of neutrons. A neutron can decay to produce a proton and an electron. Neutrons and protons then combine to form the central nuclei of atoms, with electrons subsequently captured in surrounding orbits. Thus, starting with neutrons, one could build up atoms, one by one.

Gamow got one of his former Ph.D. students, Ralph Alpher, interested in the problem—and one of Alpher's young colleagues, Robert Herman, a spectroscopist, also brought his expertise. The small group called themselves "the Three Musketeers," committed to putting the universe to rights.

The single piece of observational evidence they had that they felt supported the "bottom up" hypothesis was the fact that astronomical observations demonstrated that the universe was made up overwhelmingly of hydrogen and helium. The heavier elements such as carbon, nitrogen, oxygen, magnesium, iron, and so forth were present in minute proportions—just what one might expect if the light elements had formed first and heavier elements gradually followed. Gamow saw the original "soup" of neutrons in the early universe producing some protons that combined singly with neutrons to produce nuclei of a heavy form of hydrogen called deuterium, and some protons that combined in pairs to produce nuclei of helium. The proportions of hydrogen and helium (plus a smidgen of deuterium) given by the Three Musketeers' calculations fitted well with observations. But what about the heavier atoms? Here there was a problem, since Gamow realized that as the universe expanded the collisions between nuclear particles would have become less frequent, and as a consequence the growth by this process of atoms heavier than helium would have ceased. But a promising start had been made, and their theory was published in 1948.

The detailed scenario went something like this. In the immediate aftermath of the "creation" (we will conjecture later about the exact nature of the creation event), the universe would have been a dense "soup" of nuclear particles bathed in a primordial fireball of radiation. Indeed, the early universe would have been dominated by radiation rather than by matter. After about the first hundredth of a second the temperature would have been a staggering 100 billion degrees—hotter than the center of any star, a temperature too extreme for the nuclear particles to combine to form atoms. Within a few minutes, however, the expanding celestial soup would have cooled to about a billion degrees, so that protons and neutrons (which are themselves made up of subnuclear particles called quarks) could combine to form nuclei of deuterium (the heavier form of hydrogen) and helium. Within the first fifteen minutes of its creation, about 25 percent of the universe by mass must have been helium nuclei, with the rest remaining as hydrogen (both normal hydrogen and a minute amount of deuterium)— a figure in line with observations of the current abundance of these elements. Several hundred thousand years of expansion were then needed before the universe had cooled sufficiently to allow electrons to combine with the hydrogen and helium nuclei to form true neutral atoms. It was during this so-called epoch of recombination that the universe would have become abruptly transparent to radiation from the primordial fireball. Now the radiation could expand freely, decoupled from the matter in the universe. The nascent universe was still almost entirely hydrogen and helium (with a minor component of deuterium). The first generation of stars and galaxies would have contained just the elements of hydrogen and helium, in proportions defined by the mixture of elements in the universe when it was only a few minutes old. The formation of the more familiar heavier elements remained a matter of some speculation.

Gamow and Alpher had developed a persuasive case for the universe starting in a single creative event. But Gamow's sense of fun got the better of him in the paper announcing their analysis. He added the name of a friend, the distinguished theoretical physicist Hans Bethe, to the paper (unbeknown to Bethe). Gamow then took enormous pleasure in getting published a paper with authors Alpher, Bethe, and Gamow. His contrived authorship was a play on the first three letters of the Greek alphabet—alpha, beta, gamma—an appropriate concatenation because alpha radioactive particles are helium nuclei and beta radioactive particles are electrons, which with gamma radiation were all constituents of the primeval universe. The fact that the "alpha/beta/gamma" paper appeared, entirely by chance, on April 1, 1948, must have given the mischievous Gamow particular pleasure.

Hidden away in a 1949 paper that Alpher wrote with Herman (one of the Three Musketeers) was a prophetic suggestion that the intense radiation produced in the inferno of the creative outburst that had kick-started the universe should still be propagating through the cosmos—although as the universe expanded this remnant radiation would have been redshifted to progressively longer and longer wavelengths and would now be a faint echo of the initial outburst. It was suggested that the remnant radiation would now be in the wavelength range of microwaves. But Gamow and Herman thought that this remnant from the outburst of creation might be far too faint to be measured; surprisingly no experimentalists sought to take up the challenge to look for it at this time.

The Three Musketeers eventually decided to go their separate ways. Gamow was fascinated by the discovery by Crick and Watson of the structure of DNA and turned his considerable intellect to the challenge of unraveling the genetic code. His young colleagues went off to work in industry. It was left to others to take their ideas forward. But now they had a firm foundation on which to build a definitive theory for an expanding universe of vast age and extent.

Not all scientists were won over by the Gamow magic. One strong skeptic was an outspoken Englishman called Fred Hoyle, working at Cambridge University. Hoyle, who had been born in Bingley, Yorkshire, on June 24, 1915, was a brilliant and imaginative theorist and, like Gamow, committed to the popularization of science (his prodigious writings included science fiction novels). The son of a wool merchant, he studied mathematics and theoretical physics at Cambridge University during the 1930s. Hoyle worked on radar research during the war, when he came across two other brilliant theorists, Herman Bondi and Thomas Gold. In their spare time the three shared their ideas on cosmology and developed the notion of continuous creation—which would become known as the "steady state" theory. The idea was put forward in two papers in 1948—one by Bondi and Gold addressing the more philosophical aspects of the theory, with a companion paper by Hoyle exploring the mathematics, published two months later.

The British Broadcasting Corporation (BBC) asked Hoyle in 1949 to present a popular series of radio talks on the nature of the universe. In these talks Hoyle was dismissive of a single epoch of creation, referring sarcastically to a "big bang." Hoyle certainly did not expect his term of derision to be adopted as the legitimate name for the Lemaître/Gamow view of creation. But the unexpected happened; the creative event of Lemaître's and Gamow's theories, and their various derivatives, has been called the big bang ever since.

Hoyle was an atheist. And the last thing he could have anticipated (even in one of his novels) was that his sarcastic title would receive mention by Pope Pius XII. The church liked the big bang, seeing it as giving scientific legitimacy to the poetic description of creation in Genesis. The pope extrapolated the big bang hypothesis: "Hence, creation took place. We say: therefore, there is a Creator. Therefore, God exists!" A new "Great Debate" had been born. It was a dangerous move by the pope, since if cosmology had subsequently disproved the big bang hypothesis the above quotation could have been turned against belief in a divine creator.

One of the reasons why Hoyle did not like an explosive creation was that the whole notion of explosion ran counter to the concept of the collapse of matter under gravity that was needed to form galaxies. If all the matter in the universe had been explosively ejected from a creation event, with the universe expanding ever since, then how could gravity have acted to form galaxies?

If Hoyle was unable to accept the big bang, then what could he put in its place? He certainly was not challenging the observational evidence of the galaxy redshifts. But he was concerned about the age discrepancies in cosmology. Hoyle wanted a universe without a beginning and without an end, a universe that had always existed and would always exist, a universe that did not have a single epoch of creation and therefore would not need a "creator"—a universe that was truly eternal. Hence Hoyle, Bondi, and Gold produced their "steady state" hypothesis as an alternative to the big bang. The steady state universe was indeed envisaged as being eternal. It had always existed. As the galaxies move apart in this universe, they surmised, there is a continuous creation of new matter to fill the void. The rate of continuous creation would be modest in the extreme. Hoyle calculated that the creation of just one atom in the course of a year in a volume equal to the U.S. Capitol building would maintain the universe at a constant density. Thus the steady state theory would explain the redshift of the galaxies within an eternal universe without a beginning and with no end, a universe that would produce new stars and galaxies from the material of continuous creation.

For a short time the world of astronomy was divided. While the big bang remained the hot favorite, the steady state theory did have a small band of dedicated disciples—including several cosmologists of particular eminence.

At this point radio observations entered the picture. Radio astronomy had had its beginnings in the experiments of a Bell Telephone Laboratories engineer, Karl Jansky, during the 1930s. Jansky was investigating the nature of radio noise, particularly that generated by thunderstorms, which interfered with radio communication. This was about the time that Hubble was attempting to perfect the redshift-distance relationship and Lemaître was postulating a primordial atom. At that time radio observations would have been the furthest thing from cosmologists' minds, and certainly Karl Jansky could never have imagined that his work would eventually lead to an accidental discovery thirty years later that would provide persuasive evidence for the reality of a single creation event.

In making his observations, Jansky noted the expected radio noise from terrestrial sources. He also reported that "radiations are received any time the antenna is directed towards some part of the Milky Way system, the greatest noise being obtained when the antenna points to the center of the system."

The postwar reemergence of radio astronomy was led by a new breed of radio engineers, trained in the radar and radio direction-finding techniques of wartime so easily adapted to radio observations of the cosmos. Radio signals are generated by electrons that are in motion. In clouds of hot gas in the cosmos, electrons are bouncing around producing what is called "thermal radio emission." Radio emission is also produced when electrons are trapped in a magnetic field; this is called "synchrotron radio emission." The first discrete celestial radio source was identified in 1946 in the constellation Cygnus. By the late 1950s catalogues were being produced of hundreds of objects in the radio sky, many of which could be identified with objects also detected in optical telescopes. Some radio sources were identified as lying within the Milky Way, and some were external galaxies. One particularly bright radio source that attracted early interest was an object called 3C48 (it was the forty-eighth object listed in the third catalogue of radio sources produced from Cambridge, England—hence the nomenclature). It was found to coincide with a faint optical starlike object, and it was surmised that 3C48 could be a "radio star." However, it displayed an indecipherable emission-line spectrum that made identification of the type of "star" impossible. Soon similar objects were found, and they were given the title "quasi-stellar radio sources" (soon popularized as "quasars"). The key to the spectral riddle of the quasars was eventually found by astronomer Maarten Schmidt at the Palomar Observatory, who recognized in the complex spectrum of a quasar called 3C273 the characteristic pattern of the emission lines of hydrogen, but redshifted to such an extent that 3C273 had to be at a vast distance. The spectra for the other quasars quickly fell into place.

It was this combination of extreme distance with relative brightness that made the quasars mysterious. To appear as bright as they did while being so far away demanded that the quasars must be superenergetic objects, emitting at least one hundred times the energy of a typical galaxy; there was much speculation about them being galaxies in the very act of formation. The mystery heightened when it was found from old photographic plates that the optical brightness of 3C273 varied significantly with a period of about ten years. Now, no object can coordinate its activity on its remote side with that on its near side in less time than it takes for light to travel across it. This implied that 3C273 was no more than ten light-years across. Other quasars displayed similar dimensions. Here was the dilemma facing astronomers: not only were the quasars superenergetic, apparently emitting at least one hundred times the light of a large galaxy, but this energy had to be concentrated in a very small fraction of the size of a typical galaxy. In the decade that followed, many hundreds of quasars were discovered, some with redshifts commensurate with 90 percent of the speed of light. If the Hubble relationship was applied, then the quasars were the most distant objects detectable, lying at the bounds of the observable universe. The idea developed that quasars were the active inner nuclei of galaxies in the earliest stages of formation, with the likelihood that their gargantuan energy emissions might just be powered by a massive black hole at their center.

One of Fred Hoyle's principal opponents regarding the steady state theory was a Cambridge University colleague, radio astronomer Martin Ryle. Ryle's observations of the radio emissions from galaxies indicated that galaxies clustered closer together as distance increased. Ryle thus concluded that galaxies were more tightly packed together in the past than they are at present, as expected in a big bang universe. In a steady state universe, the average space between galaxies would remain roughly the same over the eons, since as the older galaxies drifted apart new matter leading to new galaxies would fill the void.

Intriguingly it was Hoyle who identified one of the key missing pieces in the big bang puzzle, thus making the theory he opposed more plausible. While Gamow could produce hydrogen and helium in the right proportions in his hot big bang, heavier elements represented a problem for the theory. Where had they come from, if not the big bang? Hoyle, collaborating with two British astronomers working in the United States, the husband and wife team Geoffrey and Margaret Burbidge, and theoretical physicist Willy Fowler, showed how the heavier elements could have been formed from primordial hydrogen and helium after the big bang, in nuclear reactions taking place within stars.

Hoyle always held forthright opinions on astronomical affairs. He packed lecture theaters wherever he went—and "according to Hoyle" became a catchphrase for seemingly outrageous scientific ideas. With Hoyle there were never any half measures in rigorous intellectual debates, and he made many enemies. He set up the

Institute of Theoretical Astronomy in Cambridge, which would in time become one of the world's preeminent centers of scholarship. But he fell out with the Cambridge University authorities and left in a fit of pique. He retired to the English Lake District (and later the south coast) but remained very active across many fields of astronomy.

We will make a short diversion to consider the birth, life, and death of stars. Although some scientific terms are used here, it is not necessary to understand the details of the science to appreciate the richness and wonder of the creative events in the heavens.

Stars are formed from isolated clouds of gas and dust that accumulate in the space between the stars. Such clouds of gas and dust can be observed illuminated by starlight, and these nebulae form some of the most picturesque of astronomical objects. If an interstellar cloud of gas (predominantly hydrogen and helium) collapses under the effect of gravity, the energy of infall is converted into heat, so that the collapsing cloud soon attains an extremely high temperature—on the order of 10 million degrees. At such extreme temperatures certain nuclear fusion reactions can take place. In a newly formed star, hydrogen nuclei are "fused" together to form the heavier helium nuclei with the release of vast amounts of energy. (A helium nucleus has slightly less mass than the combined mass of the component hydrogen nuclei, and as Einstein demonstrated from special relativity in his famous $E = mc^2$ equation, the destruction of a small amount of mass creates a significant amount of energy.) The liberation of this so-called thermonuclear energy increases the pressure in the mass of the gaseous material to the point where its gravitational contraction is halted. A star is born. The young star soon settles down to the relatively stable state in which it spends most of its active life. During this long period of stability, the star's self-gravity pulling matter inward is balanced by the pressure from thermonuclear energy pushing matter out. This delicate stellar balancing act is maintained at the expense of the loss of nuclear fuel. In a star like our Sun about 655 million tons of hydrogen are transformed into about 650 million tons of helium each second. The lost mass is converted to the energy that is eventually radiated from the star's surface.

And so it is with all the stars. The loss of mass and the generation of thermonuclear energy provides the answer to the question that challenged human imagination over the millennia: "What makes the Sun and stars shine?" The energy source utilized with potentially catastrophic consequences by humans in the building of thermonuclear weapons (hydrogen bombs) is the same energy source harnessed in the central nuclear furnaces of the stars.

Although the nuclear fuel reserves of a star are enormous, they are not unlimited. When the hydrogen in the central core of a star is expended, gravity again takes control. As the core starts to contract, this again causes the star's internal temperature to increase. At about 200 million degrees, the fusion of the helium ash left over from the fusion of hydrogen can take place. Helium nuclei then fuse to form carbon and oxygen. When all the helium in the core is in its turn expended, later stages of nuclear fusion may follow involving further contraction of the core and the fusion of successively heavier elements all the way to iron, beyond which no further fusion reaction can generate energy. Thus the long-sought-after goal of the medieval alchemist, to change the elements from one to another, has indeed been achieved on the cosmic scale in the centers of stars. This is the intriguing scenario for the formation of the elements that Hoyle, the Burbidges, and Fowler came up with. The new theory (which became known as the B^2FH theory) allowed the Gamow model of an initial universe made up of hydrogen and helium to evolve to a universe with a rich diversity of elements.

But if the heavier elements such as carbon, nitrogen, oxygen, magnesium, sulfur, nickel, cobalt, and iron are formed inside stars, how are they released to the interstellar medium to contribute to the formation of new stars, planetary systems, and life-forms? The lifestyle and eventual fate of a star depends on its initial mass. Massive stars (ten to one hundred times the mass of the Sun) shine the most brilliantly but burn up their nuclear fuel reserves within a few tens of millions of years. On the other hand stars of more modest size, like the Sun, live for ten billion years or longer. Massive stars may be unstable during periods of their evolution and shed part of their outer fabric to the interstellar medium. For stars some ten to twenty times the mass of the Sun, death comes in a spectacular blaze of glory. The accelerating consumption of nuclear fuel leads first to the expansion of the massive star into what astronomers call a "red supergiant." Eventually, with all its nuclear fuel reserves expended, the central core of the star collapses catastrophically to form a compact stellar remnant called a "neutron star" (which may be observable as a rapidly rotating radio beacon—a so-called pulsar). For more massive cores gravitational collapse produces the ultimate state of compaction—a black hole. The collapse of the core is accompanied by an explosive ejection of the star's outer envelope, witnessed as a supernova explosion. The energy released in these spectacular displays of celestial pyrotechnics is almost beyond comprehension; it is equivalent to the simultaneous explosion of ten billion billion billion 10-megaton hydrogen bombs! The extreme conditions of these acts of stellar suicide enrich the interstellar medium with elements heavier than iron, including platinum, silver, gold, and uranium. Such elements are rare (and therefore on Earth are deemed to be particularly precious) because the events that produce them, supernovae, are rare.

The aftereffects of a supernova explosion can be witnessed for hundreds of thousands of years as an expanding "supernova remnant." Some of the most spectacular nebulae in the sky, such as the famous Crab Nebula, are the remnants of ancient supernovae. Supernovae are also believed to produce "cosmic rays"—energetic particles permeating the whole of space, which can be detected entering Earth's atmosphere.

Supernovae disperse elements that were formed inside stars to the interstellar medium, where these elements contribute to a subsequent generation of stars that will be richer in heavy elements than their predecessors. It is a sobering thought that the atoms that make up everything of our experience, including ourselves, were processed through the central furnaces of stars eons ago. Ancient mythologies relating humans to the stars thus contained a semblance of truth—we are, in a very real sense, "children of the stars."

In stars of more modest size than those that evolve to become supernovae, when hydrogen reserves are expended the burning of helium swells the star to become a red giant. The gravity of a small star, however, is insufficient to drive it to later stages of nuclear fusion. When its helium reserves are expended the star contracts and cools to become what is called a "white dwarf." This peaceful form of death is believed to be the eventual destiny of 999 of every 1,000 stars. Our Sun is believed to be almost five billion years old. In about another five billion years its hydrogen reserves will be expended, and it will start burning helium. Its increased energy will boil off Earth's atmosphere and oceans. During this helium-burning phase, the Sun will gradually swell to become a red giant, engulfing the planets Mercury, Venus, Earth, and Mars in its expansion before gradually contracting into a geriatric white dwarf.

The theories for stellar evolution, and the big bang theory, seemed to fit reasonably well the observed abundance of the elements—at least at the level of hypothesis. But sadly cosmology was pretty short of hard observational data, and the big bang sure wasn't an event that one could repeat to test one's predictions. And besides, all was not well with redshifts and distance estimates. The estimated "apparent age" of the universe from the Hubble constant, and that for the age of the Earth derived from radioactivity measurements, remained too close for comfort (even if one chose to ignore the globular cluster data). A healthy increase in the inferred age of the universe was needed to ensure plenty of time for previous generations of stars and the progressive evolution of the heavy elements to provide Earthbound radioactive material. It would certainly be useful if the cosmological age estimate could be shown to be in error by a significant margin.

Edwin Hubble could not have been blessed with a better scientific heir. Allan Sandage, born June 18, 1926, had been enrolled as a Ph.D. student at the Californian Institute of Technology (Caltech) in 1948. Astronomy had been a boyhood passion, and it was the fulfillment of a dream to be enrolled in the first Ph.D. program at Caltech. The new program was designed to ensure effective use of the new 200-inch telescope. In 1949 Hubble was looking for an assistant to help him interpret the photographic plates of faint galaxies, and it was suggested to Sandage that he should check out what Hubble was after.

It turned out that Hubble wanted someone to help him count the number of galaxies in different parts of the sky to see what they would tell him about the way the universe was evolving. This had been a long-standing and massive program of Hubble's from the 1920s and 1930s. The basic argument was that if counting galaxies indicated that they increased in number in proportion to the surveyed volume, with no indication of an edge as with stars in the Milky Way, they could be used to map the structure of the universe. Counting galaxies and estimating their brightness (their apparent magnitude) would have a further use. Just as using the brightest star in a galaxy as a proxy measure of its distance had been of some use (so long as confusion with multiple-star systems and gaseous nebulae was avoided), so taking the brightest galaxy in a cluster of galaxies as a proxy measure of distance could be useful for clusters of galaxies at distances so extreme, and having luminosities so faint, that other methods (such as taking spectra to determine redshifts) could not easily be used.

The brightness of galaxies was a further rung in the ladder of overlapping techniques used to measure distances in the universe — a ladder whose first rung was parallax, whose second was Cepheids, whose third was bright stars in galaxies, whose fourth was redshifts, and whose fifth was bright galaxies in clusters. These various rungs in the ladder were complemented by other techniques to cross-check them and provide overlapping calibration, but these five rungs were the ones requiring the greatest care and attention. Supernovae were poised to make an especially big impact. Get the "inches" right, and the accuracy of the measurement of "miles" will follow; get the "light-years" right, and the accuracy of the measurement of "megaparsecs" will follow. That was the approach that was needed.

Sandage's joining Hubble was the start of a fruitful, albeit brief, partnership. When Hubble died in September 1953, Sandage resolved to take over responsibility for the pursuit of a firm value for the Hubble constant. He saw it as his professional destiny to extend the Hubble legacy.

Sandage, at the time of Hubble's death only twenty-eight years old and the youngest astronomer in the Caltech team, decided to go right back to fundamentals and check what Hubble and Humason had done in their earliest work with the 100-inch telescope on Mount Wilson. But now Sandage had access to the giant 200-inch telescope on Palomar Mountain, and a new generation of sensitive photographic emulsions made it possible to search out much fainter objects with spectacularly better resolution than previously available. And when he tested Hubble's use of bright stars as distance indicators in spiral galaxies too faint to allow detection of Cepheids, he was in for a nasty surprise. It turned out that many of the so-called stars used for distance calibration were not stars at all. They were bright gaseous nebulae (referred to as HII regions, being the designation for ionized hydrogen gas). Thus a simple extrapolation of distance by going from galaxies where both Cepheids and the brightest stars could be cross-calibrated, to galaxies where only bright stars were used, proved to be highly unreliable in a large number of cases. The galaxies where Hubble had inadvertently identified clouds of gas, rather than bright stars, were removed from the sample. They were replaced with those where the brightest stars had unambiguously been identified. As a result Sandage found that the Hubble constant was halved at a stroke, and the Hubble time as an age estimate of the universe was immediately increased to over six billion years. This was a most pleasing drift in the right direction. But Sandage realized that his task had barely begun, and he dedicated a lifetime in astronomy to meticulously checking and cross-checking each step of the extragalactic distance scale: from Cepheids of both populations, and other forms of variable stars, to bright stars; from bright stars to redshifts; and from redshifts to bright galaxies in clusters of galaxies. He then tried to improve the estimate of the Hubble constant and hence of the Hubble time as an indicator of the age of the universe.

By 1957 Sandage had moved the Hubble constant down to just 75 kilometers per second per megaparsec, giving an "age" for the universe of a startling thirteen billion years—comfortably longer than the best estimates for the age of the Earth, and comparable with the estimates for the age of the Milky Way derived from globular cluster stars.

Sandage teamed up with Swiss astronomer Gustav Tammann, and together they pushed the envelope even further. Tammann studied the Pinwheel Nebula, M101, in greater detail than ever before. Despite his great care in searching for Cepheids, he found none, thereby pushing the M101 distance out beyond 23 million light-years. The Hubble constant just kept on shrinking, and by 1975 Sandage and Tammann were proposing a value as low as 55 kilometers per second per megaparsec (although other teams of astronomers on the trail of the Hubble constant challenged such a low value.) A clustering of results from various distance-estimating techniques now ranged from 50 to 100, with a mean value of 75 kilometers per second per megaparsec.

While parallax observations for nearby stars (including many Cepheids) were what anchored the cosmic distance scale, later steps depended on the use of "standard candles," or of indirect methods. Cepheids (of both kinds) remained of great value as standard candles. But sometimes they were too faint to be identified in distant galaxies. The brightest stars in a galaxy were also of use (although retaining some uncertainties), but astronomers had to be satisfied that they were looking at individual stars. However, there was another, and far brighter, form of standard candle that would prove of particular value: the supernova.

With positive proof that the Andromeda spiral nebula was truly a distant and distinct galaxy, the first irrefutable "island universe" beyond the Milky Way, the brilliant Andromeda outburst of 1885, which at its peak was so bright as to outshine all other stars in the galaxy, took on a great significance. Galaxies could clearly bring forth stellar explosions of previously unimagined proportions. Walter Baade and a Caltech colleague, the Swiss physicist Fritz Zwicky, took up the challenge of looking for supernovae in other galaxies.

Zwicky was one of the more eccentric scientists of the twentieth century. He had been born in 1898 in Bulgaria but had been bought up and educated in Switzerland. Although he spent his professional life in the United States, he always retained his Swiss citizenship. He was an outspoken genius. His outspokenness meant he had few supporters in the scientific community, and many of his more ingenious contributions to research were overlooked because colleagues did not like the way he presented his ideas. Few of the history books of science refer to Zwicky, since he lacked the band of supporters and acolytes that his eminent colleagues attracted, followers who would later record their mentor's achievements. Somehow the good-natured Baade managed to cope with his eccentric colleague's extremes of behavior, recognizing his undoubted genius. They collaborated well during the 1930s on supernova studies, but sadly some ill-considered remarks from Zwicky about Baade's German origins meant that they fell out professionally and personally with the advent of war.

Zwicky was a disciple of the method of "directed intuition," the so-called many-sided or morphological method of research. While this led him to some absurd hypotheses, it did allow Zwicky and Baade to explain the mechanism of supernovae explosions, involving the collapse of stellar cores and the formation of neutron stars—thirty years before neutron stars were eventually identified with the discovery of pulsars. They also concluded (correctly) that cosmic rays, the energetic particles continuously bombarding Earth's upper atmosphere, had their origins in supernovae eons ago. Zwicky also made some startling predictions about "dark matter" in the universe, which we will have occasion to look at in the next chapter.

Zwicky started his systematic survey for supernovae in 1934, using a small 3.5-inch camera to photograph the rich Virgo cluster of galaxies from the roof of the Robinson astrophysics building at Caltech. The experiment failed to discover a single supernova during a two-year period, although Zwicky had predicted the occurrence of several during that interval. His lack of success amused the local astronomers, who were pleased to see their outspoken colleague embarrassed by his unexpected failure. Zwicky commissioned a new 18-inch survey telescope in 1936. He was determined, as he later explained, to show the professional astronomers what a determined physicist can do. Beating the "tar" out of the sky I found my first supernova in March 1937. . . . On August 26 1937 I discovered my second. . . . A third excellent one I found on September 9 1937, at the end of the same observing period.

Since the pioneering work of Zwicky and Baade, subsequent surveys have demonstrated that supernovae occur in spiral galaxies typically once every few decades. Many hundreds have been found, and much is now known about the behavior of supernovae; the death of such a star is witnessed as a brilliant outburst lasting many tens of days before fading into oblivion. Supernovae were found to fall into two principal classes—the so-called Type I supernovae (which were believed to mark the death of first-generation stars in binary systems) and the so-called Type II supernovae (which were believed to mark the death of massive later-generation stars that had evolved rather rapidly and died young). Extensive study revealed that the absolute brightness of Type I supernovae at their maximum brilliance (and especially of a subgroup called Type Ia) was remarkably similar from one to another, whereas Type II supernovae varied somewhat in their absolute brightness at maximum brilliance. Thus the Type Ia supernovae came to be seen as admirable standard candles for estimating cosmic distances.

Sandage meanwhile led a growing band of astronomers producing increasingly improved distance estimates to galaxies. Improved distances meant that the estimates for the Hubble constant and the inferred age for the universe were gradually starting to make real sense. But the elusive "smoking gun" for the big bang had yet to be found: the remnant radiation from the initial creative outburst. And until that was found, Sandage could measure redshifts until the cows came home; the big bang theory would be secured only with such direct observational evidence.

In 1964 two Bell Telephone Laboratories scientists, Arno Penzias and Robert Wilson, were hard at work on another problem. Cosmology could not have been further from their minds. The Bell company owned a strange radio antenna on Crawford Hill at Holmdel, New Jersey. This was in the shape of a large open horn, and had been designed for telephone communications using the *Echo* satellite. Communications technology had moved on, and the horn antenna was no longer required for its original purpose. Penzias and Wilson decided to try to use it for radio astronomy at microwave wavelengths, and in particular to look for faint radio emissions originating from the Milky Way—in fact from the tenuous interstellar material lying away from the plane of the Milky Way. Such observations would provide a background limit for communication satellites. In surveying radio emissions from the Milky Way, they were following in the footsteps of their illustrious Bell forefather, Karl Jansky. The Bell Labs, although focused on applied research problems, also allowed their scientists the opportunity to undertake more speculative, "pure" research—in part as a means of keeping connections with academic researchers, but also in the expectation that applied solutions might serendipitously emerge from pure research projects. History has demonstrated the benefits of this enlightened approach.

Microwave radio emissions from the Milky Way were expected to be very faint, and the extremely sensitive horn antenna was well suited to this task. However, it remained critically important to try to eliminate any spurious electrical noise that might originate from within the antenna or receiving equipment and drown out faint emissions from the Milky Way. Despite all their efforts, the horn picked up a faint residual background noise. The intensity of the unexpected noise was uniform, regardless of the direction the antenna was pointed. If the unexpected noise was cosmic in origin, then it must have come from a very large volume of the universe—not just from within the Milky Way. This seemed such an unlikely proposition to the two scientists that they immediately suspected their equipment. Was there an unexplained source of internal noise? They checked out the equipment again and again. Two pigeons roosting in the horn appeared to be the culprits; they were captured and released some distance away. However, their homing instincts meant they returned to the horn. They had to be killed, and the horn disassembled, and any evidence of their presence cleaned off. Despite all these efforts, the unexplained background noise remained, suggesting that it must indeed be cosmic in origin.

This was so unexpected that Penzias and Wilson held back from publishing their result to spare themselves embarrassment in case the noise was an artifact of their equipment or their observing method.

Any sort of body that has a temperature above absolute zero (the coldest any object can be) emits radiation due to the thermal motion of electrons within the body. The absolute zero of temperature is −273 degrees Celsius. In fact temperatures are expressed with respect to absolute zero on the so-called Kelvin scale, such that absolute zero is 0 degrees Kelvin, the freezing point of water is 273 degrees Kelvin (whereas it is 0 degrees Celsius), and the boiling point of water is 373 degrees Kelvin (whereas it is 100 degrees Celsius). Radio astronomers find it convenient to use a system in which a source of emission can be described in terms of an "equivalent temperature" of the perfectly radiating body that would emit that form of radiation. Penzias and Wilson estimated that the emission they were detecting had an "equivalent temperature" of about 3.5 degrees Kelvin.

Not far away, at Princeton University, the physicist Robert Dicke had, unbeknown to Penzias and Wilson, started building equipment designed to search for the residual or "background" radiation from the big bang. The Three Musketeers had, of course, predicted such emission—with a present equivalent temperature of 5 degrees Kelvin. Other theoreticians had followed up this prediction, but surprisingly no experimentalists had taken up the challenge to search for it prior to Dicke. And one of Dicke's young colleagues, the theoretician Jim Peebles, had started making real progress in refining the theory. Peebles argued that the early universe must have been bathed in intense X-rays, since otherwise in the first few minutes after the big bang nuclear reactions would have proceeded at such a rate that all the hydrogen would have been converted to heavier nuclei. But the fact that hydrogen today makes up almost 75 percent of the matter in the universe suggested to Peebles that the early universe must have been dominated by radiation, with X-rays blasting helium nuclei apart almost as fast as they were formed. Peebles's back-to-basics approach had come up with a figure of 10 degrees Kelvin for the present temperature of the cosmic background radiation, which would have been red-shifted over the billions of years of expansion of the universe to microwave wavelengths. (It needs to be stressed that the equivalent temperature recorded by the antenna of Penzias and Wilson is not the temperature of the present universe but rather the temperature that the universe had long ago when it became transparent to radiation, originally a very hot temperature that was subsequently reduced in proportion to the enormous expansion that has taken place over billions of years.)

Penzias heard about the Princeton work merely by chance. He just happened to be talking to a friend—who had just learned from a colleague of an interesting talk that Peebles had given explaining his prediction. The friend suggested that Peebles's work might just be relevant and perhaps Penzias should make contact with the Princeton team.

Robert Dicke had made excellent progress in designing and constructing a simple system for looking for the microwave echo from the big bang. Since the radiation would be expected to be the same from all directions, he saw no need to design an antenna with directional properties. He planned a simple, albeit extremely sensitive, system to be mounted on the roof of the Palmer Physics Laboratory at Princeton. But before Dicke and his collaborators could complete their measurements, he received a telephone call from Penzias.

It looked like the Princeton team had been scooped. But Penzias and Wilson nobly suggested that the Bell Labs and Princeton teams should publish a pair of companion letters in the prestigious *Astrophysical Journal*, which they did in 1966. Penzias and Wilson's paper mentioned the detection of the 3.5-degree background radiation without considering its cosmological implications, instead noting: "a possible explanation for the observed excess noise temperature is the one given by Dicke et al. in a companion letter in this issue." Neither of these papers mentioned the predictions of the Three Musketeers, apparently much to their irritation. But no offense had been intended; after sixteen years, most astronomers had forgotten about the prophetic work of Gamow and his young companions. In all subsequent work their pioneering theory has been given its deserved acclaim.

The discovery of a universal background microwave radiation commensurate with the big bang theory—the faint echo of creation—seemed to have served a death notice on the steady state theory (although perhaps overhastily). And it would win Penzias and Wilson the Nobel Prize in 1978 (the Nobel committee having finally come to terms with including astronomy within the physics prize).

In Rome, Georges Lemaître learned of the detection of the faint echo of the big bang just a week before he died, on June 20, 1966. Lemaître had once described the big bang as "a day without yesterday." Perhaps he might have called it "the instant before eternity."

The amiable Walter Baade retired in 1958 and returned to his native Germany in 1959. He died the following year. His genius and manifold contributions to astronomy were recognized when, in 2001, a giant new 6.5-meter telescope built on an isolated peak high in the Andes was named after him. His eccentric colleague Fritz Zwicky died in 1974, his magnificent intellectual contributions having been largely unrecognized during his lifetime because of the antagonism he had generated among colleagues.

George Gamow, the person who first provided theoretical legitimacy to the big bang, retired from the George Washington University in 1956 and moved to the University of Colorado. He stayed there until his final illness in 1968. Good-natured and humorous until the last, he acknowledged in his final days his lifestyle of fun, good company, and carefree drinking: "finally my liver is presenting the bill."

In 1983 Willy Fowler received a Nobel Prize for his part in the collaboration with Hoyle on the formation of elements within stars. Many felt it was a great injustice that Hoyle did not share in this recognition. However, at the time Hoyle was once again embroiled in controversy. He was pushing the notion that life had originated in the depths of space and that biotic material from space still brought diseases to Earth. To right the injustice of his not getting the Nobel Prize, in 1997 the Swedish Academy awarded Hoyle the prestigious Crafoord Prize in recognition of his outstanding achievements. Controversial, stimulating, more often right than wrong, and deliberately outrageous until the end, Fred Hoyle died August 20, 2001.

The twenty years on either side of the Second World War had seen contributions from giants of science, the likes of which humankind had rarely experienced before.

Chapter 6

LIVING WITH INFLATION

By the 1970s most of the remaining major uncertainties surrounding the big bang theory for the evolution of the universe had been resolved, and the theory had gained the overwhelming support of the world of science. It was taught in astronomy courses as the perceived wisdom, endorsed with the authority of the scientific establishment. The Church of Rome was happy with it (although some Protestant denominations were not). Schoolchildren learned about it. The world of the media pretended to understand it. This was no longer the stuff of controversy. Those few scientists who still sought to challenge the fundamental concept of a big bang, and the cosmological origin of the redshifts, were assigned to the fringes of the astronomical community; mainstream researchers would no longer take their proposed alternatives seriously.

The steady state theory slipped into obscurity. It was referred to only by scientific sociologists in the context of an interesting bit of history—an astronomical "debate" that never seriously developed into an issue of real consequence. Compared with the Great Debate of 1920, the big bang versus the steady state controversy was mild indeed. The cosmic microwave background had been taken as the final nail in the coffin of the steady state; most astronomers believed that it should now be laid to rest so that serious cosmologists could refine the details of the big bang.

The big bang and the steady state debate in some ways echoed that between the ideas of Anaximander and Anaxagoras from two and a half millennia earlier. Anaxagoras had envisaged that at one time "all things were together" and that the motive force for the universe originated in a single point—an imaginative idea that would come to fruition in the big bang theory. Anaximander on the other hand wanted a universe determined by "the infinite" and needed an "eternal motion" to explain the balancing process of things coming into being and passing away in an eternal universe—a steady state theory in embryo. Of course Anaxagoras and Anaximander were merely applying philosophical reason to the possible nature of the cosmos, without any observational evidence on which to base their speculations. But it is nevertheless of interest that ancient philosophy was debating the alternatives of a creation event starting the universe from a single point versus a continuous creation in an eternal universe.

In 1970s cosmology the big bang ruled supreme. No one even thought seriously of alternative theories. Of course responsible cosmologists were cautious about reading too much into their observations in terms of the nature of a "creation event." What the observations were really telling them was how the universe appeared to be evolving from its very earliest stages. (The microwave background indicated the state of the universe at the "epoch of recombination," some 300,000 years after the creation event, when the universe became transparent to radiation; cosmologists could not look further back in time.) The observations did not really reveal any details on how the whole thing had begun, what had triggered the big bang, the nature of its happening, and what had gone before.

The evidence in support of the big bang cosmology was, nevertheless, now extremely impressive, and was based on five key sets of observational results:

1. The light from more distant galaxies, having traveled for a greater time, is redshifted more than the light from nearer sources; and galaxy counts increase with redshift. Both of these observations are just as would be expected if space itself is expanding. For relatively nearby galaxies the velocity of recession inferred from redshifts is likely to be proportional to distance, but galaxies at extreme distances would be expected to depart from this linear relationship if the pace of expansion

of the universe has been changing (either accelerating or decelerating) or if space is strongly curved.

2. The expansion of space appears to have had little influence on the size of individual galaxies or the size of individual clusters of galaxies. It is the space between them that is expanding. However, in using the brightness of galaxies to infer distances care needs to be taken, since distant galaxies are seen as they were at an earlier time, when they were younger than nearby galaxies, and the evolutionary properties of galaxies may change their brightness.

3. We are immersed in a sea of faint microwave radiation, as expected if the universe used to be hotter and denser following a big bang (with the remnant radiation redshifted from X-rays to microwaves by the universal expansion).

4. The amount of helium in the universe is too great to have been created in all the generations of stars since galaxies were formed, but the relative amounts of hydrogen, helium, and deuterium are compatible with their formation in the minutes following the big bang. The creation and distribution of the elements heavier than helium can be explained in terms of nuclear processing within stars and supernova outbursts over the eons.

5. Galaxies billions of light-years distant (e.g. intrinsically brilliant quasars, probably powered by the flow of gas and stars into a central massive black hole) look distinctly younger than those that lie nearby, which is what one would expect in a universe that is evolving as it expands. (In the steady state theory one would expect to see in the nearby universe some young galaxies in addition to older, more evolved galaxies.)

All this information seemed to accord with a big bang universe expanding from a creation event (albeit one of uncertain nature). But none of the observational evidence actually told cosmologists anything about the creation event itself. For that they were locked into theory, speculation, and imagination. There was certainly no shortage of any of those. With no hard evidence on the nature of the universe prior to its becoming transparent to radiation at the epoch of recombination, and no established physics theories that could deal with the extreme temperatures and densities of the embryonic universe, you might think that the cosmologists would have given up speculating about the creation event. Quite the reverse! In the event cosmological speculation went into overdrive, despite the shortage of observational "facts" on which to base serious theories. As Mark Twain once observed: "There is something fascinating about science. One gets such wholesale returns of conjecture out of such a trifling investment of fact."

But all was still not entirely well in big bang paradise. Despite all the evidence appearing to support the theory, there were three issues that cosmologists struggled to explain. The first difficulty was that the universe appeared to be "flat"—by which is meant it seemed to sit precariously on the dividing line between eternal expansion and eventual collapse. Now while flatness accorded with the Einstein–de Sitter model, it was difficult to explain when taken alongside all the other known parameters. A "flat" universe is one where parallel lines never meet. The space curvature of general relativity does, enigmatically, allow parallel lines to meet. Think of the analogy of Earth's lines of longitude that are parallel on a flat map but then meet at the poles when we give them curvature— wrapped around the spherical surface of a globe. Or think of a saddle-type curvature for space where parallel lines can be made to diverge (if one drew parallel lines on a sheet of rubber, then stretched the rubber over the curvature of the seat of a saddle, then the lines would diverge). Once one allows space curvature, then anything is possible, and the conventional wisdom of Euclidean geometry, as learned at school, needs to be set aside. But space curvature depends on the total amount of material present in the universe, and that is the key factor determining whether it will expand forever or eventually collapse back on itself. Was the fact that some observational data suggested that the universe is flat just a lucky break? It certainly needed an explanation.

The second difficulty was the so-called horizon problem. In terms of its structure (for example as demonstrated by the distribution of galaxies, the structure of clusters of galaxies, or the uniform temperature of the microwave background) the universe is remarkably similar regardless where it is observed from and what direction you look in the sky. We describe the structure of the universe as being homogeneous (a uniform mix) and isotropic (the same in all directions) on the large scale.

But how do opposite "horizons" of the universe know how to keep in step (that is, to preserve the homogeneous and isotropic properties)? The time since the big bang has not been long enough for light to get to the farthest reaches and back (and we noted previously that an object, in this case the universe, cannot coordinate its activities on one side with those on its other in a time less than its takes for light to traverse it). It has been suggested that perhaps the speed of light was very much greater in the early universe than it is now, so that remote regions of the nascent cosmos could keep in touch. But with the invariance of the speed of light being at the heart of relativity, there aren't many physicists willing to put their money on the variable speed of light hypothesis. Even so the idea remains popular with a few cosmologists.

The third difficulty was related to uncertainty about how galaxies were formed. As Fred Hoyle had noted, the whole idea of explosive expansion runs counter to the notion of the localized collapse under gravity needed to form the first generation of galaxies. If the post–big bang expansion of space pulled all the material of the universe apart, how come some of it could then come back together again to collapse under the action of gravity to form galaxies of stars? There must have been something in the nature of the big bang that sowed the "seeds" of galaxy formation. But what could that have been?

Astronomers just hate having unanswered issues like the "flatness," the "horizon," and the galaxy "seeds" mysteries. If the big bang was the ultimate theory for the evolution of the universe, then questions surrounding the flatness, horizon, and galaxy formation difficulties had to be answered. Only then could one speculate with any degree of real scientific conviction about the nature of the actual creation event itself.

Flatness depends on the density of matter in the universe—that is, how much matter is out there per unit volume in the vastness of the cosmos? The expansion of the universe is essentially a battle to the death between gravity, pulling matter together, and the kinetic energy (the energy of motion) of the expansion. The density of matter determines the overall role of gravity in shaping the universe (remember the axiom "space tells matter how to move—and matter tells space how to curve"). Think of the analogy of a rocket being launched into space. The rocket has to attain an "escape velocity" to break free of Earth's gravitational hold on it. If the rocket is traveling faster than Earth's escape velocity, then its kinetic energy wins the battle with gravity. If the rocket fails to attain Earth's escape velocity, then gravity triumphs and the rocket crashes back to the ground.

The Greek letter omega is used to describe the so-called flatness parameter (or density parameter) of the universe. The density parameter is defined in such a way that if space is exactly "flat," then omega is equal to one. Three possible fates might be thought to await an expanding universe, with the deciding factor being the density parameter (arising from the sum of all matter in the universe producing the gravitational shaping of space). Below a critical density, equivalent to only about three hydrogen atoms per cubic meter, there is insufficient material in the totality of space for gravity to hold the universe together, and it will expand forever. In this state the universe is said to be "open." For densities greater than the critical value, gravity is sufficient to eventually bring the universal expansion to a halt and then reverse it, so that the universe will go into a contraction ending in a "big crunch." In this state the universe is said to be "closed." (In a variant of the closed universe model, a big crunch precipitates a new big bang, leading to a new epoch of expansion. Such an "oscillating universe" has no true beginning and no eventual end but merely follows a series of successive expansions and contractions punctuated with big bangs.) For a density at precisely the critical value, the universe is balanced delicately between the states of expansion and contraction. In this state the universe is said to be "flat." Omega can be defined as the actual density of the universe divided by the critical density, so that exactly at the critical density omega has the value of one. An open universe would represent a victory for kinetic energy; a closed universe would represent a victory for gravity; a flat universe would represent a draw in the battle between the two competing forces.

The fact that the universe appears to possess properties indicating that it is flat, so that the actual density must be close to the critical density, does seem somewhat strange. Nevertheless, there is reason to believe that the universe must have been born with omega exactly equal to one—since the theories predict that if omega had initially been less than one, then it would rapidly tend toward zero as the universe ages, and if it had initially been greater than one, then it would rapidly get bigger, tending toward a value of infinity as the universe ages. The fact that it appears to be even remotely close to the value one at the present time, as far as we can tell, suggests that the universe must have been born with this exact value because, since the big bang billions of years ago, there has been plenty of time for omega to have been driven to either zero or infinity had it had any other initial value than unity. A consequence of this assertion is that there must be much more matter in the universe than we can see, because in terms of visible matter (that is, the matter we can observe in stars, nebulae, and galaxies) omega appears to fall well short of the value one at the present time. Cosmological detectives are faced with this baffling "missing mass mystery."

The first hint that there was invisible matter in the Milky Way was articulated by the great Dutch astronomer Jan Oort in 1932, who concluded that visible stars near the Sun could account for only about half the mass implied by the velocities he observed of stars moving out of the plane of the Milky Way. In 1933 the controversial Fritz Zwicky entered the fray and tried to convince his ever skeptical colleagues that there must be "dark matter" in the universe. He noted that clusters of galaxies did not seem to have sufficient mass in them, at least as could be implied from their visible stars and gas, to keep them bound by gravity within a cluster. The galaxies within clusters should have flown apart, Zwicky asserted, if their total mass was contained in what you could see through an optical telescope. There really was something very odd happening. Zwicky concluded that the visible material that could be observed in galaxies within the clusters accounted for only 10 percent, or perhaps even less, of the mass needed to bind the clusters gravitationally. There was another observation that troubled the ever inquisitive Zwicky. He was puzzled by the fact that galaxies rotated in an unexpected manner. If a spiral galaxy was viewed edge-on, then one rim was observed to be blueshifted and the opposite one redshifted when the effect of cosmological redshift was removed; that is, one rim was moving toward the observer and the other away, just as one would expect if the spiral galaxy was rotating grandly in space. (Of course the galaxies are so enormous that even with the very large rotational velocities that have been measured they must take several hundred million years to complete a rotation.) No surprises so far, and astronomers had used the Doppler technique to map the rotational velocities of galaxies in some detail. But Zwicky was perplexed by the strange fact that all the stars in a galaxy seemed to be rotating at the same speed. One would have expected the stars near the center of a galaxy, where the mass of a galaxy appears to be concentrated, to be revolving around its center faster than those farther out, in the same way that the innermost planets of the solar system are revolving faster than the outer planets. In the galaxies he had studied, the stars were all revolving in tandem as if each galaxy was a solid body rather than an assembly of independent stars. Thus Zwicky decided that most of the mass of the universe must consist of "dark matter" clumped in unseen "halos" surrounding the galaxies. The revolution of these halos would drag the galaxies' visible stars around with them as if the galaxies were solid bodies, and on a larger scale the halo material would also provide the extra mass required to keep clusters of galaxies gravitationally bound together.

As might have been expected, other scientists largely ignored Zwicky's "dark matter" hypothesis. They did not like his hectoring manner, his arrogance, or his method of directed intuition whereby, when standard theories failed to explain observations, he demanded new theories and new observing methods (often of a most unexpected kind). He explained his method of directed intuition in astronomy thus:

There awaited us an unknown buried multitude of hidden treasures. . . . New cosmic bodies and phenomena which could only be divined through systematically directed intuition, and subsequent tenacious search with proper instruments.

At this time other astronomers might not have believed in dark matter, supernovae, or neutron stars, but Zwicky did, and he intended to find them. History would prove that Zwicky was correct on many of the key issues in astronomy. Unfortunately his abrasive style meant that he failed to convince fellow scientists of the validity of many of his hypotheses.

Dark matter remained in the "science fiction" category for many decades, but as observational data improved, the issue of the missing mass—required to explain the rotation of galaxies and their clustering—just would not go away. Theorists eventually started taking the issue more seriously, and once they did they proposed several novel ideas for the possible origin and nature of dark matter. Explanations for the nature of the invisible mass ranged from strange subatomic particles to small stellar or planet-size bodies.

It has been suggested that there could be some new form of fundamental particle, never detected on Earth but which the big bang produced in profusion. Such a particle or particles would have to have mass (to contribute to gravitational pull) but would interact only weakly with ordinary atoms, so that their presence could not be detected in normal conditions. These hypothetical particles were given the name WIMPs—standing for "Weakly Interacting Massive Particles." WIMPs, if they exist, would be relatively slow moving and would contribute what is called "cold dark matter" to the missing mass problem. Theoretical physicists are convinced that WIMPs, allowed for by their theories, will one day be found experimentally—although exactly how remains a matter for speculation. The supposition that they are weakly interacting means they will be extremely difficult to detect.

There is also the possibility of "hot dark matter" formed in the big bang. This could be composed of strange will-o'-the-wisp subatomic particles called neutrinos. Neutrinos have actually been detected, albeit only with advanced technology. They are formed as a product of the radioactive decay that emits beta particles and are generated in the Sun and other stars. The big bang theory predicts the creation of as many neutrinos as photons—and it is estimated that the photons in the big bang outnumbered the total neutrons and protons formed (the so-called baryons) about a billion times. Until the 1980s neutrinos were thought to have zero mass. But now it is suspected that they have a minute but certainly nonzero mass, and their accumulative effect is thought to contribute to resolving the missing mass mystery. Experiments are being carried out with sensitive detectors buried deep underground to measure the numbers of elusive neutrinos bombarding Earth from the cosmos. The experiments are carried out in deep mines to avoid contamination from background cosmic rays, since the ethereal neutrinos—unlike other forms of cosmic radiation—can pass straight through Earth's crust.

The alternative theory to subatomic particles providing the missing mass is that large planet-size bodies (about the size of Jupiter and thus not massive enough to ignite as stars) concentrate in the halos of galaxies. Or perhaps additionally a profusion of small and extremely faint stars called "brown dwarfs" might exist unobserved (because of their faintness) in the halos. These hypothetical halo objects were given the generic name MACHOs, for "Massive Astrophysical Compact Halo Objects." (If the Milky Way is over 90 percent invisible matter, as Zwicky first suggested, then it would require a population of some five trillion MACHOs—spectacularly more than the few hundred billion stars in the Milky Way.)

So it was the MACHOs versus the WIMPs when it came to describing the possible origin and nature of the missing mass. If there is indeed very significant halo material in galaxies (either WIMP or MACHO, or perhaps even both), then there should be evidence of this: light approaching us from a distant background object such as a galaxy should be bent by an intervening massive halo. It will be recalled that the bending of light in an intense gravitational field, as predicted by Einstein's general relativity, was demonstrated by Eddington's 1919 solar eclipse expedition. The bending of light by any kind of mass distorting space can produce what is called a "gravitational lens"—in the same manner as a glass lens bends light by refraction. Gravitational lens effects have indeed been discovered in a few tens of cases, caused by the invisible massive halos of intervening galaxies bending the light from more distant galaxies. In fact gravitational lenses have been invaluable in demonstrating to any remaining skeptics that the greater the redshift of a galaxy then indeed the greater its distance, since in gravitational lenses the redshift of the background galaxy (usually a quasar) is always greater than the redshift of the intervening galaxy whose invisible halo is producing the lens effect.

Dark matter certainly exists in some form or forms and helps to flatten the universe. But a problem remains. Even the most optimistic estimates of the cosmic abundance of normal visible matter (making up stars and nebulae and so forth) plus the baryon form of dark matter (MACHOs) suggest that together they can contribute no more than 4 percent of the "stuff" that flattens the universe. WIMPs might contribute up to a further 25 percent of what is needed. Hence all the visible material in the universe, plus MACHOs, plus WIMPs—all of it together just isn't up to the job. There is still a further 70 percent of "something" to be found, and theorists have started talking about "dark energy" to make up the difference. (Remember the equivalence of mass and energy, given by $E = mc^2$. If mass cannot be found, then why not search for energy?)

And now we will look at the horizon problem, in its way just as peculiar as the missing mass mystery. In 1981 the MIT cosmologist Alan Guth introduced the intriguing concept of "inflation." It needs to be stressed that the experimental evidence for inflation is scant indeed. But the circumstantial evidence that is being accumulated in favor of inflation is starting to look persuasive. Inflation proposes that in some minute fraction of a second during the creation event, the embryonic universe expanded exponentially (at faster than the speed of light), inflating the size of what is now the observable universe from smaller than even a proton out to perhaps a little larger than a grapefruit. It was in the minute fraction of a second of inflation that Guth says the future behavior of the universe was determined, after which conventional expansion began.

Guth argues that the universe was made "flat," homogeneous, and isotropic by the process of inflation.

Inflation is a rather difficult concept to understand, but it is even more difficult to appreciate what it tells us about the universe. Because everything is attracted to everything else by gravity, gravity can be thought of as acting as "negative energy." But Einstein's $E = mc^2$ equation indicates that mass is a concentrated form of positive energy. Thus it appears that one could have the positive energy in mass exactly balanced by the negative energy of gravitation, so that the overall energy could be zero; such are the strange consequences of Einstein's theories of relativity. George Gamow recalled how, in the 1940s, he was out walking with Albert Einstein and pointed out to him that according to his equations a star could be created out of nothing at all. The negative gravitational energy could be exactly canceled out by the positive energy of the star's mass, giving an overall energy of zero. Gamow noted that as the implications of this bizarre proposal dawned on him, "Einstein stopped in his tracks—and since we were crossing a street several cars had to stop to avoid running us down." The life of the greatest scientist of the age could have come to a sudden and premature end because of the sense of disbelief in his own theories.

We do not need to delve too deeply here into quantum physics, the theories necessary to move from the everyday world of visible matter to the strange world of atomic particles. Quantum physics was the handiwork of the brilliant theoretician Max Planck, in 1900. Along with Einstein's relativity, quantum theory produced the new revolution in physics. Quantum physics helps to explain the strange subatomic world, while relativity seeks to describe the large-scale structure of the universe; attempts to unify the two theories into a single worldview able to cope with the infinitesimal and the infinite have proved problematic for science. An important concept within quantum theory is Heisenberg's "uncertainty principle," which states that in the subatomic world it is impossible to say, at the same instant, both where a particle actually is and wh its momentum is. One can talk only in terms of the *probable* l tion and behavior of subatomic particles. Gamow's assertio a state of apparent "nothingness" can be separated into en gravity (negative gravitational energy, which can exactly cancel out positive energy in mass, giving a zero net energy) is a consequence of the uncertainty principle. In the subatomic world the uncertainty principle allows a particle to "tunnel" through a barrier, since its position is actually indeterminate. Sometimes a particle might find itself unexpectedly on the other side of what was thought an impenetrable barrier. (This quantum tunneling effect is not just a theoretical trick but is actually used in certain electronic devices.) The barrier need not be physical—in principle it can be an energy barrier, between "nothing" and a mixture of mass-energy and gravitational energy that exactly cancel out.

Quantum physics does allow the temporary creation of a bubble of energy out of nothing, so long as it then disappears almost immediately. The less energy there is in the bubble, then the longer it can survive. Now, as Gamow speculated, if a quantum bubble's gravitational energy and mass-energy exactly cancel out, then its zero net energy could allow the bubble to exist forever. It was just this sort of thinking that Alan Guth applied to the origin of the universe and inflation, suggesting that the universe originated from "nothing" in what he called "the ultimate free lunch." But if a quantum bubble existed containing all the mass-energy of the universe, even for an instant, gravitation would have snuffed it out immediately. So that is why Guth needed to push the limits of his imagination to propose a good heady burst of "inflation" to allow the embryonic universe to expand almost instantaneously before gravity could take hold. Gravity as we know it is an attractive force. But Guth suggested that in the universe's initial phase, gravity acted instantaneously as a repulsive force, so that it was gravity that initially pushed the embryonic universe apart. (By way of a trivial and certainly not physically valid analogy, imagine a spring-loaded jack-in-the-box. The spring initially pushes the lid open when the catch is released but then holds "Jack" back from going farther than the spring allows.)

Inflation could solve the horizon problem, since during the
⁻e fraction of a second in which inflation took place in the nas-
universe there would be time for light to crisscross such a mi-
lume, making both sides of the minicosmos aware of each

other. The exponential growth of inflation had been exactly what de Sitter's model of 1917 had predicted, attracting much ridicule at the time. His model universe expanded exponentially when a single atom was fed into it. A quantum bubble of size less than a billion-billionth the size of a proton could inflate in a minute fraction of a second to 15 centimeters (the grapefruit universe), a volume expansion of mind-boggling proportions (about a trillion trillion trillion trillion trillion trillion trillion trillion trillionfold!), leaving the hot fireball to take over driving the expansion of the observable universe.

One of the reasons that inflation is difficult to understand is that it appears to allow things to travel faster than the speed of light. But this is not strictly true. During the epoch of inflation the expansion of space is exponential, and matter is pulled along in the expansion—but *within* the inflation of space nothing is traveling faster than light.

If the big bang arose from some form of quantum bubble with rapid inflation, what existed before the big bang? Some would argue that this is a meaningless question for science, since the big bang defined when time itself began; thinking beyond the beginning of time is a matter for theologians rather than scientists. (Alternatively we could say that the big bang acts as a reference from which we could conveniently define the beginning of time, even if there was something, of which we can know nothing, that went before.) But there is a further development in inflation theory, called chaotic inflation, that suggests that a quantum bubble in a preexisting universe inflated to produce our universe—which in turn can produce bubbles that inflate new universes in a form of "budding" from our universe (which itself was "budded" from an earlier one). Thus this form of celestial horticulture offers no beginning and no end—and returns us to the concept of the "infinite" presented by Anaximander and so loved by the ancient Greeks.

Chaotic inflation has some echoes of Fred Hoyle's steady state theory, with the concept of continuous creation in a universe with no beginning and no end. However, the "young Turks" of cosmology had been brought up on the reality of the big bang in 1981 Alan Guth was asked about the steady state theory he asked, "What's the steady state theory?"

So far so good. But if inflation flattened the universe and gets rid of the horizon problem, we still have the third problem of how galaxies were formed in a smooth universe. The nascent universe did require some structure, so that gravity could allow galaxies to form. Where were the "seeds" from which galaxies could grow? Well, quantum physics can help. Within the quantum world there are density fluctuations that continually change in an entirely chaotic fashion. In the instant of creation, inflation could have blown up these quantum fluctuations and preserved them in the nascent universe at the time that inflation ended. And once the fluctuations were there, then the rest was preordained. The "seeds" of higher density were there for gravity to take control, producing a first generation of supermassive stars that could cluster under gravity to form the first generation of galaxies. These galaxies themselves would then over time be drawn by gravity to form clusters of galaxies, as well as the superclusters of clusters that now define the large-scale structure of the cosmos. Within galaxies small density fluctuations would continue to produce stars. The rest, as they say, is "cosmic history."

There is actually a way to test this idea through observation. If the original quantum fluctuations were real, then surely we should see them preserved as "ripples" in the microwave background radiation, which provides us with the earliest look we can have of the newborn universe. Well, the ripples just happen to be there. A NASA satellite called COBE, for *Cosmic Background Explorer*, was launched in November 1989 to explore the microwave background. COBE confirmed the presence of ripples, although only faintly: the background was uniform to within one part in 100,000. But such faint ripples could have seeded the formation of galaxies and everything else we now see in the heavens. Some real hints of the grand feast that represented the "ultimate free lunch" were present in the COBE data and would be demonstrated more force- ly by a later NASA mission.

e real "smoking gun" for inflation would be the detection of onary gravity waves." When a stone is thrown into a pond, the disturbance it creates flows out from the point of impact. Like-wise astronomers speculate that the sudden disruption of a gravita-tional field (for example caused by the collapse of a star's core un-der gravity to form a black hole) would be seen as ripples in the gravitational field. Attempts to detect gravitational waves here on Earth, for example those triggered by a supernova, have not been successful so far; but such "ripples" in gravity would be minute and would require detectors of remarkable sensitivity. Scientists con-tinue to build gravity wave detectors of increasing sensitivity in attempts to detect the subtle variations in gravity caused by cata-strophic events such as supernovae.

Theory does suggest that the rapid inflation of the embryonic universe should have produced gravitational waves. Gravitational waves, being disturbances in the geometry of space itself, should travel unimpeded through material that would otherwise absorb all forms of electromagnetic waves such as X-rays. As we know, up until the epoch of recombination the universe was opaque to elec-tromagnetic radiation. But if gravitational waves were triggered by inflation, they would have propagated through the dense "soup" of the early universe without trouble. Thus if inflation was a reality the subtle ripples of gravitational radiation should still be present in the cosmic microwave radiation seen today.

Observations such as gravitational lensing and cosmic back-ground fluctuations supported the idea of dark matter and the model of a big bang with inflation. But observational cosmology still had some surprises in store. For decades astronomers had been trying to identify any slowing down of the universe's expansion caused by gravity. They speculated that there could be some evi-dence hidden away in distance estimates for galaxies that might just show that the universe was expanding more rapidly in the distant past than it is at the present time. With omega equal to one, the ex-pansion would gradually slow without ever coming to a halt. But there were problems with measuring deceleration. The uncertain-ties in the distance estimates were still just too great, so that wh~ a deceleration might actually be hidden away among the data Sa~ age and others were accumulating, it couldn't be extracted enough precision to be convincing. But it was worth trying for the deceleration using what was believed to be the best kind of standard candle—the Type Ia supernovae.

As noted earlier supernovae come in various kinds, the two principal ones being the so-called Type II and Type Ia, which can be distinguished by their spectra and by the way their light varies with time (their sudden brightening followed by their dimming over days and months—this variation in brightness being called "the light curve"). As explained earlier, Type II supernovae are believed to represent the death of very massive stars, typically larger than about ten times the mass of the Sun, that have expended all their nuclear fuel reserves so that the inner core collapses under gravity to form a neutron star. This collapse is accompanied by the explosive ejection of the outer shell of the star witnessed as a Type II supernova event. Type II supernovae cannot be used as standard candles since their intrinsic brightness varies over far too large a range (presumably because the mass of the exploding stars can vary widely.) Type Ia supernovae, on the other hand, make excellent standard candles. They are thought to originate in binary star systems where one star has already evolved to become a white dwarf. The gravitational field of the white dwarf sucks in material from the outer atmosphere of its companion. A massive runaway thermonuclear firestorm, witnessed as a Type Ia supernova, is triggered at a certain stage when the mass sucked onto the surface of the white dwarf exceeds a critical value. Since the critical value that precipitates the explosion must be unique, then the spectra and light curves of Type Ia supernovae look remarkably similar. Thus when they shine forth at their brightest, Type Ia supernovae can be used as standard candles. In turns out that some Type Ias that fade more rapidly than others are actually intrinsically slightly less bright, but this effect can be corrected for. Indeed, with this correction, the inherent brightness of Type Ia supernovae can be determined with a margin of error that experts believe is less than 10 percent—which for standard candles is deemed to be exceptionally good. A Type Ia supernova brightens over a period of two three weeks and then declines in brightness over a period of hs. It is important to detect any supernova while it is brightit is to be used as a standard candle. Type Ia supernovae are the best standard candles known to astronomy. Type Ia supernovae are a million times brighter than Cepheids, so they can be used as standard candles out to very much greater distances.

Although supernovae occur in any one galaxy only once every several decades, they can be detected over a vast range of distances by searching clusters of galaxies with sensitive automated surveying techniques. A supernova bright enough to be detected explodes somewhere in the cosmos every few seconds. By monitoring thousands of galaxies in any given cluster for a month, an observer should expect to find at least one supernova somewhere in the cluster during that time.

Both Robert Kirshner of the Harvard College Observatory and Saul Perlmutter of the Lawrence Berkeley National Laboratory in California have used supernovae to look for the expected deceleration of the expansion of the universe. Perlmutter's team was the first into the race but made a slow start as the scientists perfected their equipment and data analysis techniques. Eventually, however, they were detecting ten to twenty new supernovae each search campaign. Although Kirshner's team started later, its members caught up to their California rivals rapidly. Astronomy can be a competitive business, and a healthy sense of rivalry can be good for science if a sense of urgency and clarity is brought to the "hunt." Kirshner and his collaborators called themselves the "High Z team" (Z is a symbol used for redshift); Perlmutter's team called themselves the "Supernova Cosmology Project." Each team photographed clusters of galaxies with sensitive detectors called charge-coupled devices or CCDs (just the sort of detectors, developed for astronomy and military applications, that are now used in video cameras and in digital cameras). CCDs have revolutionized astronomy from the ground and from space over the past twenty years, and they have proved to be invaluable for the photometry and spectroscopy of very faint objects. They are extremely sensitive and count individual photons from faint sources of light; indeed the are so sensitive that a distant cluster of, say, five thousand gala can be imaged in just ten minutes.

CCD images of clusters of galaxies are taken separated t days, and one image is subtracted from the other electr there are no changes, then observers will not see anything on the subtracted image. In contrast, if a supernova has blazed forth, then it can easily be detected in the subtracted image. Powerful computers are necessary to number-crunch the data as quickly as possible and to ensure definitive supernova detections. Once a supernova is found, there is a rush to other telescopes to make follow-up observations. Its variation of brightness with time is obtained, as are spectra, so that as much information as possible about the supernova and its redshift can be determined.

Once both competing teams reached their stride, each could essentially deliver supernovae on demand and measure their light curves and redshifts. The Kirshner team got a large number of detections from observations made with the Hubble Space Telescope. The rivalry to detect the expected deceleration of the universe grew in intensity.

But when the two teams independently analyzed their results, they were in for a shock. Neither found the expected deceleration; nor was it obvious that the expansion of the observable universe was entirely uniform. What the data in fact showed was that the Type Ia supernovae at large redshifts were all about 25 percent dimmer than would have been expected. It seemed that cosmology had another serious difficulty on its hands.

What could possibly explain the dimness problem? The two teams explored every possible explanation. Perhaps some form of intergalactic dust was filtering out light from the distant supernovae, making them look fainter than they really were for their redshifts. But this explanation did not stack up, since intergalactic dust of the required density should show up in the color of the supernovae—and it didn't. Blue light from a supernova would be scattered more by any intergalactic dust than would red light, so that a supernova at extreme distance would look redder than usual; but this was not the case. Perhaps the light from the distant supernovae was being spread by some form of gravitational lensing effect. But this explanation was also untenable. If space displayed se-negative curvature, then light from a distant supernova could out over a greater surface in a given time and therefore inter. However, plenty of other evidence suggested that space was flat. Perhaps distant supernovae were somehow inherently different. Suppose that in young galaxies at extreme distance stars evolved in a different way, so that their supernovae were inherently fainter than those in more evolved galaxies nearby. But while this explanation could not be completely discounted, the uniformity of the spectra and light curves of Type Ia supernovae meant that it was an unlikely way out of the dilemma.

Could it be that the expansion of the universe is accelerating? This would mean that the nearby universe, seen as it was comparatively recently, is expanding more rapidly than the universe at vast distance, seen as it was in its relative infancy. If a Type Ia supernova was farther away than its redshift implied, then obviously it would look fainter. The most obvious explanation to account for a smaller redshift than expected for a given distance would be that the universe was expanding more slowly in the past than it is now.

What could be causing this unexpected acceleration? Theoreticians returned to the possible behavior of the hypothetical "dark energy." The speculation was that perfectly empty space would produce a "vacuum energy" out of its perfect void. This "dark energy" would still allow the universe to retain its "flatness" if dark energy made up 70 percent or so of the mass-energy of the universe alongside the approximately 30 percent present as luminous and dark matter (thereby keeping omega at one.) The dark energy was envisaged as being a form of "repulsive gravity," a pale imitation of the force that had kicked off the inflation of the embryonic universe. In fact the proposal that completely empty space could produce a "vacuum energy" was not an entirely new idea. Pairs of subatomic particles, comprising a particle and its antimatter partner, do pop out of a vacuum for infinitesimal periods. Such virtual pairs of particles can indeed be detected by the energy they release. In the right circumstances such vacuum energy just might produce the negative gravitational effect required—although calculation suggested that this explanation for the vacuum energy would produce a force that was way too high and that there must be a way to attenuate it. The search for the elusive nature of the energy goes on.

The acceleration of the universe can be explained as an interpretation of Einstein's much-maligned cosmological constant. Einstein had adopted the cosmological constant as a mathematical trick to stop the universe from collapsing catastrophically under the action of its own gravitational field (since at that stage astronomers could not envisage the universe as expanding). In essence his cosmological constant was a form of "antigravity" whose influence increased with distance so that the universe would not collapse on itself under the action of its own conventional gravity. If Einstein had allowed his cosmological constant to be slightly larger, it would have enabled the universe not only to expand but also to accelerate.

An alternative hypothesis to the cosmological constant and dark energy idea has been produced invoking a new kind of physics. The authors of this hypothesis were Paul Steinhardt, Rahul Dave, and Robert Caldwell, at the time all at the University of Pennsylvania. They invoked a field that gravitationally repels, in the same way that an electrical field or a magnetic field can both attract and repel. They called this hypothetical field "quintessence," after the "fifth element" that Aristotle claimed pervaded all space and influenced everything that was. It has to be said that "quintessence" has as yet no secure theoretical basis in modern physics (although that might come); but a name (albeit one borrowed from antiquity) is a start, and the quintessence idea has attracted much interest.

Although no one would claim that inflation, dark matter, dark energy, and acceleration are a "done deal," by the late 1990s they had collectively provided a satisfactory explanation for several observed phenomena. Firstly, there was the long-standing embarrassment that the age of the oldest stars seemed to be extraordinarily high and too close for comfort to the "age" of the universe implied by the Hubble constant. But if the universe was expanding more sedately in its distant past, then its age would be greater than inferred from measuring its current expansion rate from the Hubble constant, and all was well again. Certainly there would have been cient time for the evolution of the elements in stars and for the ferred from stars in globular clusters of about twelve billion be correct. Secondly, we had a satisfactory explanation for the brightness of supernovae, both near and far. Thirdly, we had a ready-made explanation for the gravitational lens phenomena, whereby halos of dark matter surrounding intervening galaxies bend the light from objects at extreme distances. And finally, the origin of the small fluctuations in the microwave background, observed by the COBE satellite, had a very natural explanation in quantum fluctuations from the embryonic universe being preserved in the inflationary expansion phase.

There was still the issue of the Hubble constant. By the 1970s Sandage no longer had the field to himself—and by the 1980s there was still a disturbing disparity between the value of 55 kilometers per second per megaparsec that he and Tammann were advocating and values of up to 100 kilometers per second per megaparsec that others argued for with equal passion. It was felt that the Hubble Space Telescope, launched in 1990, would resolve the issue, and a "Key Project Team" of collaborating scientists was formed for this purpose. Unfortunately, they had to wait until the Hubble Space Telescope was given "spectacles" to correct its poor eyesight. When, after launch, astronomers checked out the new view of the heavens, they expected to see beautifully sharp images reflecting the perfect optics of the telescope and the fact that from space the distorting effects produced by small disturbances in the atmosphere (which usually cause star images to "twinkle") would be absent. They were in for a major disappointment. Instead of nice crisp images, the stars looked like hollow disks. The mirror of the Hubble Space Telescope was supposed to be the most perfect ever produced, but sadly the measuring technique used to check the shape and quality of the surface while it was being polished was flawed. NASA immediately launched an inquiry, and made plans to fit correcting optics to the telescope to fix the distortions. The busy Space Shuttle schedule meant, however, that it was not until December 1993, three years after the initial launch, that astronaut returned to the Hubble Space Telescope to do a repair job in spa

The scientists on the Key Project Team first turned their a tion to the galaxy M100 in the Virgo cluster. They found a haul of Cepheids (which could never have been detec the ground) whose careful analysis provided a distance

57 million light-years. This was the most distant galaxy for which Cepheids had ever been detected; the Hubble Space Telescope was starting to show its real power. The researchers studied about two dozen galaxies in similar detail, and by 1999 they were convinced that they had finally nailed down the value of the Hubble constant with a precision previously unachievable.

On May 25, 1999, NASA called a press conference. This would be a big day for cosmology, offering no less than the best-ever estimate for the Hubble constant. The members of the Key Project Team announced their value: it was 70 kilometers per second per megaparsec, with a precision of plus or minus less than 10 percent. This gave the Hubble time, the apparent age of the universe, as fourteen billion years.

Sandage and Tammann were not pleased. Having invested more time and effort into the quest for the definitive value of the Hubble constant than anyone else, they were promoting a value of 55 kilometers per second per megaparsec (implying a Hubble time of eighteen billion years.) But the Key Project Team's value gained instant respectability. At last the majority of scientists felt comfortable with the apparent age estimate for the cosmos derived from the Hubble constant of fourteen billion years, along with twelve billion years for its oldest stars and four and a half billion years for the age of the Earth. All the age estimates and distance estimates derived from different techniques were forming what looked to be a consistent set—at last. And since, when margins of error were included, the vast majority of estimates of the Hubble constant were consistent with the Key Project Team's value, then 70 kilometers per second per megaparsec, plus or minus 10 percent, was going to keep most astronomers happy.

In June 2001 NASA launched a successor mission to its highly successful COBE satellite, the *Microwave Anisotropy Probe* (MAP). The mission's results, when announced in February 2003, con-ned all the outcomes from COBE and improved on even the ble estimates for the age of the universe. The MAP data im-hat the universe is "flat" and made up of 73 percent dark ll the MAP results were consistent with the inflationary the big bang.

But what does acceleration mean for the true age of the cosmos? Just how old can the universe be if it was expanding more slowly in the past? And presumably as the density of the material in the universe falls as the cosmos continues to expand, then the effect of dark energy generated in the void will increase, so that the acceleration will increase with time.

Perhaps the evolution of the universe went something like this. The expansion of the early universe, post inflation, was dominated by gravity. The density of the material was such that gravity did gradually decelerate the expansion (as Perlmutter and Kirshner had still expected to witness) But as the universe expanded and the density fell, dark energy came more and more into play. However, gravity remained the dominant influence for at least seven or eight billion years. Perhaps some five billion years ago, about the time the solar system was formed, the gradually emerging effect of dark energy surpassed the influence of gravity so that dark energy became the dominant force (to the point today where it represents over 70 percent of the mass-energy of the cosmos). The expansion of the universe thus changed from a state of deceleration to the state of subtle and gradually increasing acceleration now suggested by the supernova data.

Does this all sound vaguely reminiscent of the steady state theory? Hoyle and colleagues proposed that as the universe expands and its density falls, new matter is created to fill the void. What the new cosmology proposes is that as the universe expands and its density falls, new dark energy is created to fill the void (keeping the universe "flat" and driving its expansion).

Now cosmologists speculate that at some point in the far distant future dark energy may rapidly decay so that gravity does eventually triumph, and if they are correct then there will indeed be a big crunch, bringing forth a new big bang universe—again returning to the concept that there is no "beginning" and no ultimate "end" but rather a chaotic form of eternity. Again there are themes in scenario reminiscent of the eternity of the steady state.

After the bitter debate between the big bang and stead proponents, it is somewhat amusing that both theori to have contained a semblance of truth—just as did of Shapley and Curtis in the Great Debate of 1920. It is through such disagreements that science progresses and our understanding of nature advances. We are still some way from a full understanding of the nature, the origin, and the ultimate fate of the universe; but thanks to great debates and disagreements between creative individuals the path to an ultimate understanding is being defined.

The delicate interplay of dark matter and dark energy in a post-inflation cosmos would appear to explain everything we now observe. The best estimate for the age of the universe in this scenario is on the order of thirteen and a half billion years.

So are all the mysteries of the cosmos, its size, mass, and age, now fully understood? Well, no. But fewer pieces of the puzzle are missing, and the pace of understanding is increasing. Science is now on the verge of answering many of the questions that have intrigued humankind since antiquity.

Science has come a long way in establishing the size of the cosmos, through establishing the various steps in the distance measurement ladder. Astronomers have been able to probe the remote regions of a vast and expanding cosmos, and have produced novel ideas on its likely origins and possible fate.

The first key rung on the cosmic distance measurement ladder is parallax for the nearest stars. Despite the efforts of many astronomers before and after William Herschel, astronomy had to wait until the nineteenth century for Friedrich Bessel and his rivals to determine the distance to nearby stars by the method of parallax. Even the nearest stars were much farther away than anyone had supposed, giving credence to ancient Greek and Copernican thinking that the failure to detect parallax easily was due to the vast distances to the stars.

The second rung in the cosmic distance measurement ladder is Cepheids. The work of Henrietta Leavitt and Harlow Shapley, on Cepheids, produced a new understanding of the size of the Milky Way. Shapley and Curtis couldn't agree on the nature of nebulae. Once Edwin Hubble had detected Cepheids nebula in Andromeda, however, then the issue was resolved in favor of "island universes." The true enormity of the cosmos was now open to enlightened interpretation.

Bright stars provide the third rung in the cosmic distance measurement ladder—but care needs be taken to ensure that only single stars are used. Uncertainty in this regard led Hubble and Humason to make mistakes in their classic work on distances to galaxies beyond the range of Cepheid observations.

The fourth rung, and probably the most important in finally determining the nature of an expanding universe, is redshift, which Hubble revealed was related to distance. Here was evidence of an expanding universe, introducing a new era of enlightenment on the origin and evolution of the universe.

The fifth rung in the distance ladder is the brightness of galaxies in clusters, but now increasing care must be exercised to ensure that the evolutionary behavior of galaxies is understood. The mysterious quasars suggested that young galaxies were very different from their older cousins.

And finally we can seek to use Type Ia supernovae as standard candles to explore the structure of the cosmos at great distances. As measured by these light sources it really does seem that the expansion rate of the universe was slower when it was very much younger.

The contributions of many astronomers, from the ancient Greeks to modern times, have led to the present understanding of a cosmos of a size that stretches human understanding to its limits. There have been great debates aplenty. Ideas were passed from patron to acolyte—from Thales to Anaximander to Anaximenes, from Hubble to Sandage to the host of "young Turks" of modern cosmology.

The importance of the work of philosophers and scientists over the centuries, and the excitement they have generated, can be summarized in the words of one of the ancients, Claudius Ptolemy:

> I know that I am mortal and the creature of a day; but when
> I search out the massed wheeling circles of the stars, my
> no longer touch the Earth, but side by side with Zeus
> self, I take my fill of ambrosia, the food of the gods.

In modern times Ptolemy's sense of wonder in face of the unknown was echoed by Albert Einstein:

> The fairest thing we can experience is the mysterious. It is the fundamental emotion which stands at the cradle of true art and true science. He who knows it not and can no longer wonder, no longer feel amazement, is as good as dead, a snuffed-out candle. Enough for me the mystery of the eternity of life, and the inkling of the marvellous structure of reality, together with the single-hearted endeavour to comprehend a portion, be it ever so tiny, of the reason that manifests itself in nature.

This sense of wonder and awe for the majesty of nature, professed so eloquently by two philosophers exploring the cosmos more than eighteen centuries apart, demonstrates the bond between all those seeking to understand the marvels of creation.

GLOSSARY

absolute luminosity The total energy emitted per second from any astronomical object.

absolute zero The lowest temperature theoretically possible, at which the energy of atoms and molecules is minimal. Equivalent to −273 degrees Celsius.

absorption spectrum A continuous-spectrum light source shining through a substance will display dark (absorption) lines at colors identical to those the substance would emit if it was itself emitting light.

acceleration The rate of change of velocity. Measured in meters per second squared.

accretion disk A disk of gas swirling down onto a compact star orbiting a normal star.

alpha particle A radioactive particle emitted from certain unstable atomic nuclei; contains two protons and two neutrons, so is equivalent to a helium nucleus.

Andromeda Nebula A large spiral galaxy nearest to our own, which was the first such object to be proven to be an independent "island universe."

angstrom A unit of wavelength of light, equal to one ten-billionth of a meter.

antimatter Subatomic particles that have the same mass as their normal-particle counterparts but have the opposite of some other property. For example a positron has the same mass as an electron but carries a positive rather than negative electrical charge.

apastron The farthest point of separation between two celestial bodies in orbit around each other. (In the case of the Moon or an artificial satellite in orbit around Earth, the point of maximum separation is called the apogee.)

apparent luminosity The total energy received per second per unit of receiving area from any astronomical object.

astronomical unit The mean distance between the Sun and Earth.

atmosphere The mixture of gases enveloping a planet. In the case of Earth, the main constituents of the dry atmosphere are nitrogen, oxygen, argon, and carbon dioxide.

atom The smallest component of an element that can exist and still retain the characteristic properties of the element.

atomic mass The mass of an atom, measured relative to the isotope carbon-12 having an atomic mass of 12.

atomic number The number of protons in the nucleus of an atom (equal to the number of orbiting electrons in an electrically neutral atom).

baryons A class of strongly interacting atomic particles, including protons and neutrons.

beta particle A radioactive particle (an electron) emitted from certain unstable atomic nuclei.

big bang The "event" some thirteen and a half billion years ago that marked the creation of space and the beginning of time.

binary star Two stars orbiting one another.

biosphere The whole of the land, sea, and atmosphere inhabited by living organisms.

black hole A compact cosmic object whose gravitational field is so strong that its "escape velocity" (the velocity an object must have to escape from it) exceeds the velocity of light.

blueshift The blueward shift of light from galaxies approaching ours.

brown dwarf A cosmic object intermediate between a planet and a star, not massive enough to initiate fusion.

carbon dating The isotope carbon-14 is radioactive. By measuring the ratio of this radioactive isotope to normal carbon-12 in a sample of organic material (for example wood), its age can be estimated with some precision.

centigrade (Celsius) scale The temperature scale that is defined by zero degrees as the temperature of ice and water in equilibrium and 100 degrees as the temperature of steam above boiling water.

Cepheids Stars whose brightness varies with a period dependent on their intrinsic brightness.

ters Groupings of stars or galaxies.

A fragment of debris from the formation of the solar system, orbiting the Sun in a highly elliptical orbit. When a comet is close in to the Sun, material evaporating from its surface is swept back by the solar wind into a characteristic tail.

conservation of energy The conversion of energy from one form to another, without any overall loss.

continuous spectrum An unbroken sequence of wavelengths over quite a wide range.

convection A means by which heat is transferred, whereby hot air (or some other fluid) rises and is replaced by cold air (or fluid).

cosmic rays Energetic particles of cosmic origin hitting the Earth's atmosphere.

cosmological constant A term inserted by Einstein into his general relativity equations to balance the effect of gravity and allow for a static universe.

cosmological redshift The stretching of the wavelengths of light from distant galaxies caused by the expansion of space during the time the light is in transit through the cosmos.

cosmology The study of the origin, present state, and future fate of the universe.

critical density The value of the mass density in the present universe that would bring the universal expansion to a halt.

decay In radioactivity, the spontaneous release of a particle from the nucleus of an atom.

deuterium A heavy form of hydrogen (with a neutron as well as a proton in its nucleus).

diffraction The bending of waves on passing through an aperture or on meeting the edge of a barrier.

diffraction grating A piece of glass engraved with fine parallel lines, used to form spectra by diffraction and the constructive interference of light (used in a modern spectrograph in place of a prism).

disintegration A process in which an atomic nucleus breaks up spontaneously into two or more components.

dispersion The splitting up of light into its component colors by the process of refraction.

Doppler effect The shift in wavelength observed when a source of sound (or light) is moving relative to an observer; the detected wavelength is shortened if the source is moving toward the observer and lengthened if receding from the observer.

electromagnetic waves Waves generated by changing electric or magnetic fields; visible light, gamma and X-rays, ultraviolet and radiation, and radio and microwaves are all forms of electromagnetic waves.

electron The subatomic particle carrying the unit of negative electrical charge.

element A substance that cannot be broken down into simpler substances by chemical means.

elementary particles The assortment of fundamental particles making up everything in the universe.

elliptical galaxy A type of galaxy with characteristic ellipsoidal form.

emission spectrum Light emitted by a substance in the form of discrete colors characteristic of the elements making up the substance.

energy The capacity to do work (that is, to move an object). Measured in joules.

equinox Either of the two occasions during the year when the Sun appears to cross the celestial equator (either from south to north in the "vernal" equinox or from north to south in the "autumnal" equinox).

escape velocity The minimum velocity an object must acquire to escape from the gravitational field of a celestial body.

expansion of the universe The continuing expansion of the universe from the initial big bang some thirteen and a half billion years ago; observed by the recession of galaxies and the background radiation from the big bang.

fission A breakup of certain heavy atomic nuclei stimulated by the capture of a neutron, with a release of energy and further neutrons. The basis of nuclear chain reactions.

flare A violent ejection of hot gas from the surface of the Sun or a star.

focal length The distance between the center of a lens used to focus light and the point at which parallel rays of light are brought to a focus.

focus The point at which rays of light converge.

force A "push" or a "pull"; any action that tends to alter a body's state of rest or uniform motion. Measured in newtons.

frequency The number of oscillations per second of an oscillating or vibrating object.

·sion The forcing together of light atomic nuclei to form a heavier ~ucleus, with a release of energy.

A conglomerate of billions of stars, bound by gravitational at- ɔn; sometimes found in characteristic spiral or elliptical forms in irregular configurations.

gamma rays High-energy electromagnetic radiation, emitted from the nuclei of radioactive elements.

gas The state of matter in which atoms and molecules have freedom of movement, so that a gas always fills its container regardless of its quantity.

general relativity The theory developed by Einstein relating the effects of gravity to those of accelerated motion. A prediction of the theory was that light would be deflected in a gravitational field (confirmed experimentally).

globular cluster A cluster of about a million stars. The distribution of globular clusters in the sky was taken as strong evidence that our Sun is not at the center of the Milky Way.

gravitation The universal attractive force acting between all matter.

gravitational waves Waves resulting from a disturbance in a gravitational field.

hadrons Particles that participate in strong interactions; hadrons include the baryons and mesons.

helium The second-lightest and second most abundant element in the universe.

homogeneity The state of the universe whereby its general properties appear the same to all observers wherever located.

horizon The distance beyond which no light signal would have had time to reach us.

Hubble constant The constant in the relationship between the speed of recession (as measured by redshift) and the distance of galaxies.

Hubble's law The law relating the velocity of recession of galaxies and their distance.

hydrogen The lightest and most abundant element in the universe.

infrared radiation Electromagnetic radiation of wavelengths just longer than those of visible light.

ion An atom that has acquired electric charge by the addition or subtraction of electrons.

ionization The process by which atoms lose or acquire electrons.

irradiation The exposure of material to radioactive particles or ionizing electromagnetic radiation.

irregular galaxy A conglomerate of stars with an overall irregular shape rather than the ordered elliptical or spiral forms seen in other galaxies.

isotope Different forms of the same element; the nuclei conta same number of protons but differing numbers of neutrons. For example the isotopes oxygen-16, oxygen-17, and oxygen-18 all contain eight protons, but eight, nine, and ten neutrons respectively.

isotropy The property of the universe that it looks the same in all directions.

joule The conventional unit of energy, equal to the work done when a force of one newton moves an object one meter.

kinetic energy The energy (capacity to do work) of a body in motion. Measured in joules.

light The form of electromagnetic radiation that can be detected by the human eye.

light-year The distance light travels in one year, equivalent to 9.46 trillion kilometers.

Local Group The cluster of galaxies to which the Milky Way belongs.

magnitude The brightness of an astronomical object on a logarithmic scale. "Apparent magnitude" is related to how bright an object appears to be; "absolute magnitude" relates to how bright an object actually is.

mass The amount of material contained in a body; a measure of the body's inertia. Measured in kilograms.

mass number The total number of protons plus neutrons in an atom of an element.

mesons A class of strongly interacting subatomic particles.

meteor A fragment of cosmic debris burning up in Earth's atmosphere (observed as a "shooting star").

meteorite A fragment of cosmic debris large enough to survive entry through Earth's atmosphere — typically from boulder size to tens of meters across before entry.

microwaves Electromagnetic waves of wavelengths intermediate between infrared radiation and radio waves.

Milky Way The conglomerate of 400 billion stars within which our solar system lies.

molecule The fundamental unit of a compound, made up of two or more atoms bonded together.

ula A gaseous cloud in the cosmos.

inos Elementary particles that have no charge, travel at the speed ght, and have minuscule mass.

neutrons With protons, the principal building blocks of atomic nuclei. Neutrons have a slightly greater mass than protons but do not have an electrical charge.

neutron star In the imploding core of a supernova (the event that heralds the explosive death of a massive star), protons and electrons merge to become neutrons. The surviving dense core (the neutron star, made up almost entirely of neutrons) spins rapidly and may be detected as a pulsar.

newton The fundamental unit of force, it is the force required to give a mass of one kilogram an acceleration of one meter per second squared.

Newton's laws The fundamental laws of motion: the first law states that every body remains in its state of rest or uniform motion unless acted on by an external force; the second law states that force is the product of mass times acceleration; and the third law states that for every action there is an equal and opposite reaction.

nova A thermonuclear explosion on the surface of an evolved star (probably caused by transfer of matter from the atmosphere of a companion star).

nuclear energy The energy that can be extracted from the nuclei of atoms, either through the process of fission (the splitting of massive nuclei) or through fusion (the merging of light nuclei).

nuclear fission See fission.

nuclear force The strong, short-range force that holds the particles in an atomic nucleus together.

nuclear fusion See fusion.

nucleon A proton or neutron.

occultation The passage of one celestial body in front of another, as in the occultation of a star by the Moon.

optics The study of light.

orbit The path of one object around another, whether bound by gravitational attraction or some other force.

oscillation A periodic vibration.

parallax The apparent angular displacement of a nearby star with reference to distant "background" stars as Earth orbits the Sun.

parsec A unit of cosmic distance. One parsec is the distance to a star with stellar parallax of one second of arc. One parsec is equivalent to 3.2615 light-years.

periastron The closest point of approach of two celestial bodies orbit around each other. (In the case of the Moon or an artificial satellite in orbit around Earth, the point of closest approach is called the perigee.)

period The time interval between adjacent crests of a wave, or the time to complete one cycle of a regularly repeating phenomenon.

photon The "quantum" (fundamental packet) of electromagnetic radiation.

photosphere The visible surface of the Sun.

pitch The characteristic of sound that describes its "highness" or "lowness" to a listener. Related to the frequency of the sound.

plasma A highly ionized gas.

polarization The process of restraining the oscillation of the electric field in an electromagnetic wave to a particular direction.

positron A fundamental particle with the same mass as an electron but carrying the positive unit of electrical charge. The "antiparticle" of an electron.

potential energy The capacity of an object to perform work, as a consequence of its position. Measured in joules.

power The rate at which work is done or energy is transferred. Measured in watts, equivalent to joules per second.

prism A block of glass with triangular cross section used to refract (bend) light; since different colors are refracted by different amounts, a prism will spread incident light into a spectrum of its component colors.

proper motion The shift in position of an astronomical object across the sky at right angles to the line of sight, usually measured in seconds of arc per year.

proton With neutrons, the principal building blocks of atomic nuclei. Protons have a slightly smaller mass than neutrons and carry the basic unit of positive electrical charge.

pulsar A rapidly spinning neutron star emitting a collimated beam of radio waves, which are detected as a series of rapid pulses as the star rotates.

quantum A minimum quantity by which energy can change (quanta, plural).

quantum theory The theory devised by Planck whereby energy can be emitted only in discrete amounts called quanta. The energy of quanta is given by the Planck constant times the frequency of the emitted energy.

quarks The fundamental elementary particles from which all hadrons are formed.

quasars Highly luminous celestial objects at extreme redshifts (and therefore at extreme distances, so that they are observed at an early stage in the evolution of the universe). Quasars are thought to be nascent galaxies with their phenomenal brightness powered by a massive black hole at the center.

radiation The term used to describe both electromagnetic waves and radioactive particles.

radioactivity The spontaneous emission of energetic particles (alpha particles and beta particles) and gamma rays from the nuclei of certain elements.

radio astronomy The detection of radio waves from celestial objects.

radiocarbon dating See carbon dating.

radio waves Electromagnetic waves of the longest wavelength.

red giant A large luminous star at an advanced stage of evolution.

redshift The redward shift of light from galaxies receding from ours.

reflection The return of all or part of a wave or beam of particles on encountering a boundary.

refraction The bending of a wave as it passes obliquely from one medium to another.

relativity Theories relating to relative motion; see general relativity and special relativity.

rest energy The energy of a particle at rest—the amount of energy that could be released if the total mass was annihilated, with $E = mc^2$.

scattering The process by which electromagnetic radiation is deflected by particles in the medium through which it is passing.

science The human endeavor dedicated to understanding the nature of, and patterns of behavior in, everything around us; and to making predictions based on that understanding.

scintillation The "twinkling" of stars, caused by turbulence in the atmosphere.

shock wave A narrow region of high pressure formed when a projectile passes through a fluid at a speed faster than the speed of sound.

signal The means by which information is transmitted.

solar activity The variable nature of the Sun, as evidenced by the appearance and disappearance of sunspots, solar flares, and other features.

solar system The Sun and its system of nine planets, their moons, and interplanetary objects such as asteroids and comets.

solar wind The flux of energetic particles radiated by the Sun.

special relativity The theory developed by Einstein to describe the effects of relative motion, based on the proposition that the velocity of light is independent of the velocity of its source. Special relativity led to predictions about time dilation, length contraction, and the equivalence of mass and energy.

spectrometer An instrument using either a diffraction grating or a prism to spread light into its component colors; spectrometers can also be made for other forms of electromagnetic radiation to separate the component wavelengths.

spectroscopy The study of spectra produced by a spectrometer.

spectrum A range of electromagnetic radiations spread out according to their wavelength (spectra, plural).

speed The distance covered per unit of time. Measured in meters per second.

speed of light The fundamental constant of relativity, assumed to be uniform throughout all space and time. Equal to 299,729 kilometers per second.

spiral galaxy A conglomerate of stars with the characteristic form of intertwined spiral arms.

star A self-luminous celestial body, generating energy by nuclear fusion in a central core.

steady state theory The theory developed by Hoyle, Bondi, and Gold in which the average properties of the universe never change—as the universe expands, new matter is created that fills the void and keeps the density constant.

subatomic particles The fundamental particles from which atoms are formed.

summer solstice The time at which the Sun appears to reach its highest point above the celestial equator; marks the longest day of the year.

sunspot A dark area on the Sun's photosphere, marking a region of lower than average temperature. The frequency of sunspot appearances waxes and wanes in an eleven-year cycle.

supernova The violent explosion of a massive star that has reached the end of its normal evolution.

synchrotron radiation The electromagnetic radiation produced when electrons are accelerated to near the speed of light in a magnetic field. This type of radiation occurs naturally in the cosmos, but it is also generated artificially in particle accelerators to produce intense sources of light and X-rays for probing the states of matter.

technology The devices, systems, and processes, derived from scientific knowledge and engineering practice, that contribute to our lifestyles in a useful way.

telescope An instrument able to collect light from a faint, distant object in order to produce a visible image of it; telescopes can utilize either a lens or mirror system (or a combination of both) to collect and focus the light.

terminator The line of demarcation between the illuminated and dark sides of a planet or moon.

thermal equilibrium The state in which the amount of thermal energy gained equals the amount lost.

thermodynamics The study of the laws that govern the nature of heat.

thermonuclear reaction See fusion.

ultraviolet radiation Electromagnetic radiation with wavelengths just shorter than those of visible light.

uncertainty principle The principle, defined by Heisenberg, stating that it is not possible to define both the position and velocity of an atomic particle simultaneously, since an attempt to measure one will perturb the other.

universe Everything that is known to exist.

upper atmosphere The outer reaches of Earth's atmosphere, above an altitude of about 300 kilometers.

vacuum A space from which as much gas as possible has been evacuated.

variable star A star that increases and decreases its size periodically, and changes in luminosity.

velocity The speed of a body in a specified direction.

visible spectrum The wavelengths of electromagnetic radiation that can be detected by the human eye.

watt The unit of power (energy per unit of time).

wave A periodic disturbance in a medium (for example a water wave) or in space (for example light). A wave is characterized by its amplitude, its velocity, its period, and its frequency.

wavelength The distance between adjacent crests in a wave.

weight The force with which a body is attracted to Earth. Measured in newtons.

white dwarf The residual object left after a star of comparable mass to our Sun has used all its nuclear fuel. It is believed that some 999 out of every 1,000 stars will eventually become a white dwarf.

winter solstice The time at which the Sun appears to reach its lowest point below the celestial equator; marks the shortest day of the year.

work The work done by a force acting on an object is given by the force times the distance moved by the object along the line of application of the force. Measured in joules.

X-ray astronomy The detection of X-rays from celestial objects.

X-rays Energetic electromagnetic waves with wavelengths shorter than those of ultraviolet radiation. X-rays are extensively used for medical applications.

X-ray stars Star systems that emit X-rays.

zenith The point on the celestial sphere directly above the observer.

zodiacal light A faint glow in the sky in the west after sunset, and in the east prior to sunrise, caused by the scattering of sunlight from interplanetary dust.

BIBLIOGRAPHY

CHAPTER 1: INGENIOUS VISIONS

An excellent general treatment of the foundations of astronomy is *The Cambridge Illustrated History of Astronomy* by distinguished historian Michael Hoskins (Cambridge University Press, 1997). Another excellent read on this and later periods is *The Great Copernicus Chase*, an anthology of thirty-six incidents in the history of astronomy by the eminent scholar Owen Gingerich (Cambridge University Press, 1992). In researching this chapter we made frequent reference to *Early Greek Philosophy* by Jonathan Barnes (Penguin Classics, 1987); Barnes is a scholar of international acclaim. Although over seventy years old, *Greek Astronomy* by Sir Thomas Heath (J. M. Dent, 1932) presents fascinating insights to the work of the ancients.

CHAPTER 2: SERIOUS MEASUREMENTS

Parallax: The Race to Measure the Cosmos by Alan W. Hirshfeld (W. H. Freeman, 2001) is a wonderful read. The creative atmosphere of the epoch of enlightenment is captured in *Tycho and Kepler* by Kitty Ferguson (Walker and Co., 2002), *Starry Messenger* by Peter Sis (Farrar Straus and Giroux, 1996), and our *Newton's Tyranny* (W. H. Freeman, 2001).

CHAPTER 3: THE GREAT DEBATE

The published versions of the presentations of Shapley and Curtis are contained in *The Scale of the Universe* published in the *Bulletin of the National Research Council*, volume 2, part 3, May 1921. A scholarly interpretation of the exchange of letters preceding and following the debate is contained in *The Great Debate: What Really Happened* by Michael Hoskins (*Journal of History of Astronomy*, volume 7, 1976); we drew extensively on this source. Arthur Eddington's classic *The Expanding Universe: Astronomy's Great Debate, 1900–1931* was republished by Cambridge University Press in 1988. Harlow Shapley recounted this historic period for astronomy in *Through Rugged Ways to the Stars* and *Of Stars and Men* (Greenwood Press Reprint, 1984). We had to work hard to find material on Henrietta Swan Leavitt and Annie Jump Cannon; these two wonderful astronomers are surely worthy of definitive biographies.

CHAPTER 4: SEEING RED

In 1935 Edwin Hubble presented the Silliman Lectures at Yale. These were published a year later as *The Realm of the Nebulae* (reprinted in 1982 by Yale University Press), giving the great man's personal insight to his historic discoveries. *Origins: Our Place in Hubble's Universe* by John Gribbin and Simon Goodwin (Overlook Press, 1998) is worth a look.

CHAPTER 5: THE NATURE OF CREATION

The Birth of Time by John Gribbin (Yale University Press, 2001) is an enjoyable and easy read, as is *Aeons: The Search for the Beginning of Time* by Martin Gorst (4th Estate, 2001). *Walter Baade: A Life in Astrophysics* by Donald Osterbrock (Princeton University Press, 2002) gives an interesting insight to one of the principal players; a similar tome on the enigmatic Fritz Zwicky is long overdue, and we had to pick up snippets of his intriguing insights from a variety of sources. Fred Hoyle's autobiography, *Home Is Where the Wind Blows: Chapters from a Cosmologist's Life*, presents a fascinating account of the life of this brilliant individual.

CHAPTER 6: LIVING WITH INFLATION

We have no hesitation in recommending four excellent books dealing with current views of the cosmos: *The Inflationary Universe: The Quest for a New Theory of Cosmic Origins* by Alan Guth (the genius behind inflation) and Alan Lightman (Perseus Publishing, 1998), *The Accelerating Universe: Infinite Expansion, the Cosmological Constant, and the Beauty of the Cosmos* by Mario Livio (John Wiley, 2000)—if you read only one other book on astronomy this year, make it Livio's—*The Extravagant Universe: Exploding Stars, Dark Energy, and the Accelerating Cosmos* by Robert Kirshner (Princeton University Press, 2002), and *Echo of the Big Bang* by Michael Lemonick (Princeton University Press, 2003).

INDEX

ABOUT THE AUTHORS

DR. DAVID CLARK is a New Zealander who has lived and worked in England for the past twenty-nine years. He is married to Suzanne; they have three sons and live in Oxford. He started his career as a telecommunications engineer. His research background has been in space technology and astronomy. He has published some eighty research papers in learned journals, as well as writing numerous articles for popular science and technology magazines. Until 1985 he led the Space Astronomy research team at the Rutherford Appleton Laboratory in Oxfordshire, before moving into science and technology administration. He is currently director of research and innovation at the United Kingdom's Engineering and Physical Sciences Research Council.

MATTHEW CLARK is the eldest son of David Clark. He is a first class honors graduate in classics from Balliol College, Oxford University, and received a postgraduate teaching diploma from Cambridge University. He teaches classics and ancient history at Shrewsbury School, Shropshire, England.